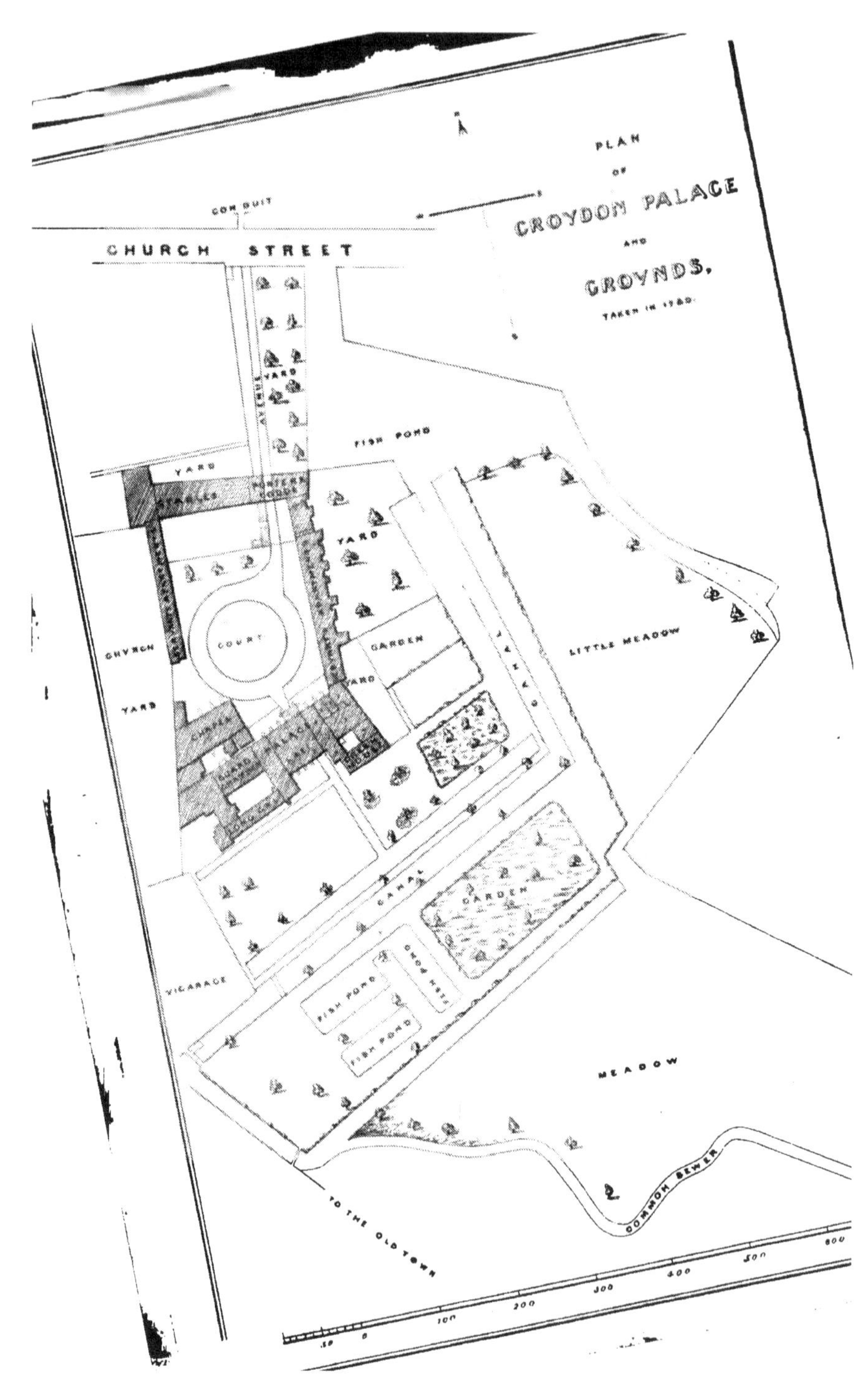
PLAN
OF
CROYDON PALACE
AND
GROVNDS,
TAKEN IN 1780.
CONDUIT
CHURCH STREET
AVENUE
YARD
FISH POND
YARD
STABLES
PORTERS LODGE
YARD
COURT
GARDEN
YARD
CHVRCH
YARD
CHAPEL
CANAL
LITTLE MEADOW
CANAL
GARDEN
VICARAGE
FISH POND
FISH POND
MEADOW
COMMON SEWER
TO THE OLD TOWN
50
0
100
200
300
400
500
600

A

HISTORY OF CROYDON:

BY

G. STEINMAN STEINMAN, ESQ., F.S.A., F.L.S.

LONDON:
LONGMAN, REES, ORME, BROWN, GREEN, & LONGMAN.

M. DCCCXXXIV.

5516

LONDON:
W. M'DOWALL, PRINTER, PEMBERTON ROW,
GOUGH SQUARE.

TO

THE MOST REVEREND

WILLIAM

LORD ARCHBISHOP OF CANTERBURY,

PRIMATE AND METROPOLITAN,

&c. &c. &c.

THIS VOLUME

IS, WITH PERMISSION, MOST HUMBLY INSCRIBED

BY HIS GRACE'S OBLIGED

AND VERY OBEDIENT SERVANT,

THE AUTHOR.

PREFACE.

It is now many years since a History of Croydon has been given to the public—for I pass over, as unworthy of notice, a modern publication, very inaccurately compiled from the works of Ducarel, and Manning and Bray. In 1783, a thin quarto appeared, from the pen of the celebrated antiquary, Andrew Coltie Ducarel, LL.D. F.R.S. and F.S.A., intitled "An Account of the Town, Church, and Archiepiscopal Palace of Croydon," forming the twelfth Number of the "Bibliotheca Topographica Britannica;" on which foundation—the earliest—every succeeding History will be raised.

Great as is the merit due to this indefatigable Antiquary, for the deep research he has displayed; still he has left much untold, his work presenting little more than the materials for a History of Croydon.

The next account of this parish will be found in Lyson's " Environs of London," (5 vols. 4to, 1792—1811); which, with that in Manning and Bray's " History and Antiquities of Surrey," (3 vols. folio, 1804—1814), comprises all that has yet been written concerning it.

The two last-mentioned Histories—if so they may be termed—are necessarily brief; they contain, however, information not to be found in Ducarel.

Of the merits, if any, of the present volume, it would ill become the Author to speak. He contents himself with stating, that he has sought diligently among the MSS. in the libraries of the British Museum, of Lambeth Palace, and of the

College of Arms; with what success he leaves the reader to determine.

THE kind assistance of his friend, SIR WILLIAM WOODS, Clarenceux, F.S.A., &c., the Author acknowledges with thanks.

G. S. S.

CAMBERWELL, *July* 18, 1833.

INDEX TO ILLUSTRATIONS.

The Illustrations are engraved by Messrs. Branston & Wright, from Sketches chiefly made by W. Sims and the Author.

CONTENTS.

APPENDIX.

INDEX TO EPITAPHS.

ADDENDA ET CORRIGENDA.

Page

32, *add, as a note,*—On the south gable of Bencham manor-house is the date 1604, formed in the brick work; by which it appears that its time of erection was early in the reign of James I., and not in that of Henry VIII., as I supposed.

49, *add, as a note,*—Sir John Tunstall must have come to Addiscombe after 1619, as, on the 13th September in that year, he subscribed his name to Alleyne's Quadripartite deed of Dulwich College, being then of Carshalton; probably Captain Poynings Heron resided here previously to Sir John.

61, *l.* 4, *for* "2,007*l.* 19*s.* 4*d.*," *read* "2,017*l.* 19*s.* 4*d.*"

62, *l.* 9, *add* "ætatis suæ 38."

109, *l.* 20, *for* "a bend of the second and third," *read* "a bend of the second. Second and third cheque," &c.

124, *l.* 15, *for* "lands," *read* "lambs."

137, *l.* 5, *after vicar's name, add* "LL.D., who occurs vicar in 1376."*

137, *l.* 6, *after* "presented," *add* "1 September, 1387."

144, *l.* 18, *for* "1684," *read* "1648."

157, *l.* 28, *for* "Grindali," *read* "Grindalli."

158, *last line, for* "bordar," *read* "bordure."

175, *l.* 14, *for* "miro," *read* "miror."

183, *l.* 6, *for* "Comb," *read* "Coombe."

189, *l.* 14, *for* "Resurrecto," *read* "Resurrectio."

l. 15, *for* "Rego," *read* "Regno"

l. 17, *for* "Januarii," *read* "Mar[ij]."

203, *l.* 14, *for* "who departed," *read* "and departed."

288, *l.* 11, *for* "30 Eliz." *read* "38 Eliz."

ADDITIONS TO REGISTER.

"1551, June 28.—Willia. Mapleton sepult. qui dedit paupibs. hujdi. pahij. p. spaim. quiqe. anno. duos denarios solvend. ad quatuo. terminos anni cora. gardianis ecclie."

* Reg. Sudbury, fol. 26 b.

"1625, July 21.—Richard Vaughan, sonne to the Lord Vaughan, and Mrs. Bridget Lloyd, were marryed."

Richard, only son of John, first Baron Vaughan and Earl of Carberry, of the kingdom of Ireland, was created a Knight of the Bath at the coronation of Charles I. He served with success on the side of royalty, during the civil wars, and held the appointment of Lieutenant-General of Carmarthen, Pembroke, and Cardigan. In this capacity he greatly distinguished himself for zeal and gallantry, and on the 25th October, 1643, was rewarded with a seat in the British Parliament, by the title of Baron Vaughan of Emlyn, in the county of Carmarthen, and after the Restoration was appointed Lord President of the Principality of Wales.

He married, first, Bridget, daughter and heir of Thomas Lloyde, Esq., of Llanlees, Cardigan, as above; second, Frances, daughter and co-heir of Sir John Altham, Knight, of Oxby, Hertfordshire; third, Lady Alice Edgerton, daughter of John Earl of Bridgewater, and died in 1687, having issue by his second wife only.

"1627, June 25.—Sir Henry Lee, Knight, father of Mrs. Smith, of Coombe, was buryed."

"1639, June 18.—Basset Cole, gentleman, and the Lady Aymie Mordant, the widow of S^{r}. Robert Mordant, Knight and Baronet, were maryed."

Lady Mordaunt was the daughter of Sir Austin Southerton, Knight, of Norfolk, and mother to Sir Charles Mordaunt, Bart.—a cavalier who paid the penalty of his loyalty by the forfeiture of his estates.

"1671, November.—Mrs. Elizabeth Cleiver, wife of William Cleiver, Doct^{r}. in Divinity, was burried in the Lady Scudamore's grave, in ye middle chancell, ye xxvth."

Whether this *lady* was Mistris Bernard, or that "*most excellent gentlewoman*, Mrs. Ream's daughter-in-law," mentioned in the case of the Inhabitants of Croydon, cannot now be determined.

ADDENDA.

Page 5, *l.* 13, *after* "about 1550." *add*, Richard Crowley, in his satirical epigrams, published, according to Ritson, in the same year, has the following:—

THE COLLIER OF CROYDON.

It is said that in Croydon there did sometyme dwell
A Collyer that did al other colyers excel,
For his riches thys collyer might have been a knight,
But in the order of knighthood he had no delight.
Would God al our knights did mind coling no more
Than thys collyer did knighting, as is sayd before;
For when none but pore collyers did with coles mell,
At a reasonable price they did their coles sell;
But synce our knight collyers have had the first sale,
We have pay'd much money, and had few sacks to tale.
A lode, that late years for a royal was sold,
Wyl cost now XVI shillings of sylver or gold.
God graunt these men grace their polling to refrayne,
Or else bryng them back to theyr old state agayne;
And especially the Colliar that at Croydon doth dwell,
For men think he is cosin to the collyar of hell."

Ibid. after l. 21, *add*—In "Damon and Pythias," a comedy written by Richard Edwards, and first acted in 1566, Grimme, the Collier of Croydon, is a character; and Ulpean Fulwel, in his comedy of "Like wil to like quod thee Devil to the Colier," published in 1568, causes three of his dramatis personæ, to wit, Tom Collier, Nichol Newfangle, and the Devil, to dance together to the tune of "Tom Collier of Croidon hath solde his cole." There is also a play called "The Historie of the Çollyer," acted before Queen Elizabeth, in 1576-7, which relates to this same worthy.

Page 16, *add as a note to line* 9—"Croinden Archiepiscopi Cantuariensis, r. c. de j marca, pro defalta: In th. b. et quieta est." Madox's History of the Exchequer, p. 384. In the Lansdowne MSS. (No. 324), I find the following entry:—"31. H. 2 Crouinedna Archipi. Cantuar. deb. 1. m. per def."—Records of Surrey and Sussex, by Le Neve. (Qu. whether this is the same fine as that mentioned by Madox.)

Page 26, *after l.* 16, *insert* "1645.—On the 15th of April the order undergiven was issued by the Parliament for the withdrawel of a detachment of 200 horse and 100 dragoons, which had till then been quarter'd in this town—there awaiting the concentration of the county forces.—*Vide* "*A Diary, or an Exact Journal,*" No. 48.

"Sir,—We have now received some intelligence that the rebells of Kent are in some measure dispersed. And therefore there being no further use of yo[r] Horse & Dragones that we wrot unto you to send towards Croydon, We desire you, they may bee recalled and disposed As you please.

Signed in the name and by the warrant of the Committee of Both Kingdomes by Your very affectionate friends and humble servants,

"*Darbie House*, MANCHESTER, LOUDONN*."
15 *April*, 1645.
S[r] THOMAS FAIRFAX."

Page 35, *l.* 15, *after* "messuage and farm," *add* "and comprises about 400 acres of arable and wood land."

Ibid. l. 19, *after Holy Trinity, add* "On an extent made in 1287, it was found that Jeffry de Haspale held certain lands for life, to be

* *MSS. Brit. Mus. Ayscough's Cat.* No. 1519, p. 44. The insurrection here alluded to was caused by a party of impressed men who had destroyed their convoy, whilst marching to join General Fairfax. They attacked and took possession of Sir Percival Hart's house at Rootham, and were immediately joined by the discontented of the county, to the number of 500 foot and horse; but Colonel Blunt being expressly sent against them, they were defeated and dispersed on the day of this order.

inherited by William and Philip de Padyndenne. Of these held in fee was the manor of Croweham."—*Esch.* 2 Ed. I, n. 25.

Further particulars of John Gage.—(Vide pp. 43, 44).

John Gage, as I learn from Bayley's "History and Antiquities of the Tower of London," was committed to that prison, by warrant bearing date 10th of January, 1590, and Beesley, or as he is there called Besseley—the cause of his misfortunes—on the 18th of December, the same year. He appears to have occupied the "Broad Arrow Tower," in which room, between the first and second recesses, on the left-hand side, is yet extant the following inscription (copied from Bayley)—apparently his work:—

QVOD RATIO REDDENDA ERIT DEO CUM
VENERIT DIES ILLA IVDICII MAGNA DE
CVNCTIS COGITATIONIBVS VERBIS ET OPE-
RIBVS. DNS ILLVMINABIT ASCONDITA TE-
NEBRARVM ET MANIFESTABIT CONSILIA
CORDIVM CVM VENERIT ..ORS. OMNE
DE VERBIS
VERBUM OTIOSVM QVOD LOCVTI FVERINT
HOMINES REDDENT RATIONEM DE EO IN
DIE JVDICII.—MAT. 12.
CVNCTA QVÆ FIVNT ADDVCET DEVS IN IVDI-
CIVM ERO OMNI ERRATO SIVE BONVM SIVE
MALVM SIT.—ECCLESIASTES, 12.

QVOTIES DIEM ILLAM CONSIDERABO TOTO CORPORE CONTREMISCO
SIVE ENIM COMEDO . SIVE BIBO . SIVE ALIQVID ALIVD FACIO . SEMPER
IN AVRIBVS MEIS SONARE VIDETVR TREMENDA ILLA VOX IVR-
GITE MORIVI VENITE AD JVDICIVM.

QVOD SIBI QVIS .. ERIT PRÆSENTIS TEMPORE VITÆ
HOC SIBI MESSIS ERIT CVM DICITVR ITE VENITE.

"The above pious memorial," says the antiquary, "is without name or date, but the characters in which it is written so closely correspond with the remains of an adjoining inscription as to leave little or no doubt of its having been made by the same person. This latter, though much defaced, appears to have been cut with surprising ingenuity, and is subscribed 'January 1591, I. Gage,'"—

a prisoner, he conjectures, of the Roman Catholic persuasion; but, of whom, he adds, "no account has hitherto been discovered." Gage appears, from the memoir of his son, to have remained in durance upwards of thirty years, and to have outlived the "remainder of his subsistence," and the several annuities "his noble allies and kindred had bestowed upon him." He left issue by his two wives (Margaret, daughter of Sir Thomas Coply, Knt., of Gatton, Surrey, and "Mrs. Barnes"); five sons—Sir Henry; Thomas, author of "Travels in Spain, &c.;" George, prothonotary in England for the See of Rome; Francis, president of the English Coll. at Douay, and John, author of "The Christian Sodality, or Catholic Hive of Bees;" and a daughter Mary: but as they were all born after his imprisonment, consequently they could not be natives of this parish.

There appears to be some confusion in the accounts given of this gentleman, from which our statements in the text are made. His attainder, and the consequent forfeiture of the manor of Haling are there mentioned; yet it is also stated, that his son relinquished the entail in favor of his father, whose life is said to have been spared by the "*King's* reprieve." It is to be remembered that he had been imprisoned eleven years previous to James' succession—an extension of time that precluded an idea that he was at that era under sentence of death. Besides his unfortunate brother Robert, I find, from the church-register, he had two others—Edward, christened 26 Aug. 1567, and William, christened 3 Oct. 1568—and a sister Mary, christened 3 Oct. 1563; which names have not yet been incorporated in any pedigree of this ancient family.

Page 62, *l.* 9, *add* "ætatis suæ 38."

Page 78, *add as a note to line* 16—"On Monday, August 11, 1635, the following entry was made by the prelate in his diary:—'One Robert Seale, of St. Alban's, came to me at Croydon and told me somewhat wildly about a vission he had at Shrovetide last, about not Preaching the word sincerely to the people; and a hand appeared unto him, and death, and a voice bid him go tell it the Metropolitan of Lambeth, and made him swear he would do so; and I believe the poor man was overgrown with fancy; so troubled myself no further with him or it.'"—(*History of the Troubles and Tryal of Abp. Laud*, 1695, p. 50).

Page 109, *l.* 8, *in note, add*—["Cheque *or* and *az.* a chief of the first."]

Page 113, *l.* 15, *after* "other of his papers," *add*—"And again this same witness, who appears to have been employed by the primate on the windows both of Croydon and Lambeth palaces, affirmed, that 'he found a picture of God the Father in a window at Croydon and Archbishop Cranmer's arms under it, and that he pulled it down;' evidence which called from Laud the remark that it had been placed there during the primacy of so zealous a prelate, as Cranmer was well known to be—and that it had been removed in the days of him against whom it was now advanced, in support of his alleged inclination to popery."—(*Hist. of the Troubles of Abp. Laud*, p. 517).

Page 137, *l.* 5, *after Vicar's name, add*—"L.L.D., who occurs vicar in 1376."—(*Reg. Sudbury*, fol. 265).

Ibid. l. 6, *after* "presented," *add*—"1 September, 1387."

Page 168, *ll.* 4 and 14, *after* "bendlets," *add*—"in sinister chf. a cross croslet."

ADDITIONS TO REGISTER.

1550, Aug.—"Mr. [Henry] Tonstall buryed the 21."

Eldest son of Sir John, by his wife Penelope, daughter of Sir Walter Leveson, Knt., of Lilleshall, Salop.

1551, June 28.—"Willia. Mapleton sepult. qui dedit paupibs. hujdi. pahij. p. spaim. quique. anno. duos denarios solvend. ad quatuo. terminos anni cora. gardianis ecclie."

1544.—"Thoms Heyrne [Heron] obijt 2 die Octobris."

1558.—"Sepulta fuit Mag^{re}. Elizabeth Heron, vidua [of Thomas Heron, Esq.] 1^{a} die Augusti."

1578.—"Richard Gornarde, the son of Bryan Gornarde, was chrystened the viij day of Marche."

Sir Richard Gornarde, or Gurney, as he is pleased to call himself, born 17th of April, 1577, was the son of Bryan Gornarde, of Croydon, a descendant of the Gurneys, of Kendall, in Westmorland, —(*Le Neve's MS. Pedigrees of Extinct Barts. Herald's Coll.*)— by his wife Magdalin, daughter of ——— Hewett. He had, I find from the parish register, two brothers—John, baptized 9 Dec. 1576, and Robert, baptized 28 May, 1681; and the name of Gornarde is frequently mentioned in this authority. In apprenticing his son to a Mr. Coleby, a silk-mercer in Cheapside, his father was singularly fortunate, as that worthy citizen at his death, bequeathed to his apprentice his shop and effects, estimated at the value of 6,000*l.*—*Lloyd.* His fortune now rapidly increased, and with it he acquired the reputation of great integrity and moderation, insomuch that he became a leading character in the city, and subsequently filled its highest offices.

On Tuesday, Nov. 25, 1641, being then Lord Mayor, he entertained the king and the royal family, on his majesty's return from Scotland, at a cost of about 4,000*l.*, when the king was pleased to confer upon him the honour of knighthood; and, on the 14th of the next month, the higher distinction of a baronetage. In this year, he caused the royal proclamation against the militia to be publicly read within the city, and was, for so doing, deprived of his mayoralty and his liberty, as we have stated in p. 25. He married, first, Elizabeth, daughter of Henry Sandford, Esq., of Berchington, in the Isle of Thanet, by whom he had, Richard, who died vita patris, Elizabeth, the wife of Thomas Lord Richardson, and Anne, who married Sir John Pettus, Knt., of Suffolk. His second wife was Elizabeth, daughter of one Richard Gosson, a London goldsmith. He was buried at St. Olive's Jury.

1582.—"Samuell Ffynche [primus] vycar of Croydon & Elizabeth Kinge the daughter of John Kinge & Clemence, were married the viij[th] day of Marche, anno dni. 1582, by Richarde Worde, Parson of Bedington, under the lycence of Edmunde Archbishop of Canterburie's grace."

1585.—"Memorandu. that the xxv[th] day of Julye word was broght to the towne of Croydon, that there lay one dead in a close nye Pollarde hill, who was putrified & stank in most horrible manner; wherefor none cold be gotten by the officers to bringe him; whereupon he lay there [till] the Tuesday at nyghte after, beinge the xxvii[th] day, at which time the Vicar [Samuel Fynche, *primus*] hired one Robert Woodwarde, & they two went unto him & found hym lyeng on his backe wth his legs pulled up to hym & his knees lying wide, his right hand lying on his right legge & his left cross his stomacke, the skin of his face & the hear [of] his hed beaten of wth the weather, no pportyon in the lineaments of his body to be proaved, they ware so putryfied, a rnt. rotten canvas dublet & his hose ragged, a blacke felt hat wth a cypres bande & two laces tyed at the ende of the band. Woodward digged the grave hard by hym where he lay & they two pulled hym in, wth each of them a large forke."

1585.—" Wm. Edsone beinge sicke (as he confessed to his wife, Willm. Andrews & one Hedd of Streatham, yet constrained hymself to goe forth to mowinge at Streetham the xxixth day of July, & comming home from thence betweene Streatham bridge & the further Norberie gate, fell downe dead, & was buried the xxvth day of Julye."

——— " Roger Pryce leaninge on a calyver charged wth hayle shotte on his left side, his matche in the same hande, the peece discharged soddenlye & kylled hyme presently, savinge as much tyme as wherein he prayed the standers by to pray to God for hym, & soe fallinge downe desiered God hartely to forgive hym all hys synnes, & soe dyed the xxvith day of Juely. And was buried the xxvijth."

1598.—" Samuell Ffynche [*primus*] Vicar of Croydon & Elsabethe Swan ware married at Sandersteed the xijth day of June."

1607, April.—" Rycharde Esteinge, a young man, beinge killed suddenlye wth a stroke of thunder & lightninge on the [neck] & under the right eare: but nothinge but blacknesse seene & the ...... of ...... swealed, was buried the xixth day: And smelt of Brimstone exceedingle."

1612.—" Elizabeth Bradberie, neece to the most Reverende ffather of famous memorie Dr John Whitegifte sometime Archb. of Cant. dieinge in London was brought to Croydon and buried in the Chappell where he lyes the xxixth day of June, anno dni. 1612, according to her request." Vide Epitaph, p. 205.

——— " M$^{rs.}$ Marie Abbot, Wydowe: Syster in Lawe to the

Archbishop of Canterbure was buried the xv[th] day of September. Anno Dni. 1624."

Daughter of John Millet, Esq., of Guilford, and widow of Anthony Abbot, Esq., of the same place, next brother to the archbishop of that name.

1625, July 21.—"Richard Vaughan, sonne to the Lord Vaughan, and Mrs. Bridget Lloyd, were marryed."

Richard, only son of John, first Baron Vaughan and Earl of Carberry, of the kingdom of Ireland, to which titles he succeeded, was created a Knight of the Bath at the coronation of Charles I. He served with success on the side of royalty, during the civil wars, and held the appointment of Lieutenant-General of Carmarthen, Pembroke, and Cardigan. In this capacity he greatly distinguished himself for zeal and gallantry, and on the 25th October, 1648, was rewarded with a seat in the British Parliament, by the title of Baron Vaughan of Emlyn, in the county of Carmarthen, and after the Restoration was appointed Lord President of the Principality of Wales.

He married, first, Bridget, daughter and heir of Thomas Lloyde, Esq., of Lanlees, Cardigan, as above; second, Frances, daughter and co-heir of Sir John Altham, Knight, of Oxby, Hertfordshire; third, Lady Alice Edgerton, daughter of John Earl of Bridgewater, and died in 1687, having issue by his second wife only.

1627, June 25.—"Sir Henry Lee, Knight, father of Mrs. Smith, of Coombe, was buryed."

1628, Dec. 6.—Sepult. "Margaret Hatton, daughter of S[ir] Robert Hatton [and niece to Sir Christopher Hatton, K.B., and Sir Thomas Hatton, Bart.]

1630, Sep.—Sepult. "The Lady Margaret Abbot."

Daughter of Bartholomew Barnes, Esq., and second wife of Alderman Sir Maurice Abbot, Knt., Lord Mayor in 1638, and fifth brother of the Archbishop.

——— Oct. 28.—Sepult. "Richard Tomerton, Keeper of the Archbishop's house."

1631, June 25th.—"William Arnold, a young man, *et magnæ spei*, was buryed."

1633, Jan. 30.—Sepult. "Ralph Smith, yeoman of the guard."

1636, Sep. 9.—Bap. "Thomas Harvy, the sonne of Mr. Eliab Harvy."

Thomas Harvey was nephew to Dr. William Harvey, the celebrated discoverer of the circulation of the blood. Eliab was an opulent Turkey merchant, the Doctor's fifth brother, and eventually the inheritor of the greater part of his effects. There are, I believe, several of this family buried at Croydon, among whom was Daniel Harvey, a merchant of London, and fourth brother of the Doctor.

1637, May.—"Christopher Heydon, the sonne of S^r. John Heydon, Knt."

Sir John Heydon, of Barkinstrop, Norfolk, was Lieutenant of the Ordnance to Charles I., from the first muster of the royal army in 1642, to his death; 16th Oct. 1653.

This Christopher is said to have been afterwards a Knight,—(*Blomefield's Hist. of Norfolk, by Parkin*, vol. vi, p. 510), but I do not find his name among the MS. lists in the Coll. of Arms.

1639, June 18.—"Basset Cole, gentleman, and the Lady Aymie Mordant, the widow of S^r. Robert Mordant, Knight and Baronet, were marryed."

Lady Mordaunt was the daughter of Sir Austin Southerton, Knight, of Norfolk, and mother to Sir Charles Mordaunt, Bart.—a cavalier who paid the penalty of his loyalty by the forfeiture of his estates.

1641, Nov.—"John Tonstall, the sonne of Henry Tonstall, Esquire, sonn and heire of S^r. John Tonstall of Adscombe, Knight, was baptized."

——— Dec. 15.—"Michael Miller, Esquire, and Barbara Astry, the daughter of S^r. Henry Astry, Knight," [of Harlington, Beds.]

1650, Aug. 3.—"S^r. Edward Parteridge's sonn buryed."

Sir Edward Parteridge, knighted 31 July, 1641, was of Langley, Kent.

1666, May 22.—Sepult. "Mr. William Austry, Curate of Croydon," [under the Rev. Dr. Clewer!]

1669, June 9.—Bap. "Mrs. Margaret Sheldon, daughter of Sir Joseph."

Sir Joseph Sheldon, Knt., was nephew to Archbp. Sheldon, and Lord Mayor in 1676.

1671, November.—"Mrs. Elizabeth Clewer, wife of William

Clewer, Doct^r^. in Divinity, was burried in the Lady Scudamore's grave, in ye middle chancell, ye xxvth."

Whether this *lady* was Mistris Bernard, or that "*most excellent gentlewoman*, Mrs. Reame's daughter-in-law," mentioned in the case of the Inhabitants of Croydon, cannot now be determined.

1674, Sept. 5.—Sepult. "Mr. John Morris, Curate."

1675, Ap. 11.—"Mr. Wm. Crow, Schoolmaster, was buried."

This is the suicide chaplain of Whitgift's Hospital, for an account of whom vide p. 70, where the time of his death is incorrectly given. Wood informs us, that the Rev. John Osborne, M.A., commenced "a Catalogue of our English writers on the Old Testament," and had printed about eight sheets when Crowe's catalogue appeared, which caused him to relinquish his design. The same author also tells us, that Crowe "took many things" from the "Felix Consortium" of Edward Leigh, when he composed his "Elenchus Scriptorum in Sacram Scripturam, &c." London, 1672; which is, questionless, "the catalogue" under a different cognomen.

RARE PLANTS FOUND AT CROYDON.

Eriophorum vaginatum—*Single-headed Cotton-grass*, or *Hares'-tail Rush.*

Agrostis spica-venti—*Silky Bent-grass.*

Centunculus minimus—*Bastard Pimpernel.*

Asperula cynanchica—*Squimancy-wort.*

Galeum anglicum—*Wall Bed-straw.*

Anagallis arvensis, γ—*Blue Pimpernel.*

Verbascum lychnitis—*White Mullein.*

Vinca minor—*Small Periwinkle.*

Chlora perfoliata—*Yellow-wort.*

Butomus umbellatus—*Flowering Rush.*

Dianthus armeria—*Deptford Pink.*

Cerostium semi-decandrum, β—*Mouse-ear Chickweed.*

——— arvense—*Field Chickweed.*

Cistus surrejanus—*Dotted-leaved Cistus.*
Mentha piperita—*Peppermint.*
Origanum vulgare—*Common Marjoram.*
Limosella aquatica—*Water Mud-wort.*
Trifolium scabrum—*Rough Trefoil.*
Hypericum montanum—*Mountain St. John's Wort.*
Prenanthes muralis—*Ivy-leaved Lettuce.*
Carduus pratensis—*Meadow Thistle.*
Phascum curvicollum—*Crooked-stalked Earth-moss.*
——— curvisetum—*Short bent-stalked Earth-moss.*
Trichostomum flexifolium—*Wave-leaved Fringe-moss.*
Tortula aristata—*Short-pointed Screw-moss.*
Campanula rapunculus—*Rampions.*
Splachnum ampullaceum—*Purple Gland-moss.*
Dianthus deltoides—*Maiden Pink.*
Genista Anglica—*Needle Furze, or Petty Whin.*
Hypericum elodes—*Marsh St. Peter's Wort.*
Rosa spinosissima—*Burnet Rose.*
Scabiosa columbaria—*Lesser Field Scabious.*
Spiræa Filipendula—*Drop-wort.*
Trifolium ochroleucum—*Yellow-flowered Trefoil.*
Triticum caninum—*Bearded Wheat-grass.*
Ophrys apifera—*Bee Orchis.*
——— muscifera—*Fly Orchis.*

CORRIGENDA.

Page.

11, *l.* 8, *for* "St. Peter," *read* "All Saints."

70, *l.* 13, *for* "of," *read* "on."

99, *l.* 4, *for* "Tentys," *read* "Fyennes."

109, *l.* 20, *for* "a bend of the second and third," *read* "a bend of the second. Second and third, &c."

127, *l.* 14. *for* "1567," *read* "1568."

131, *l.* 23, *for* "25th," *read* "27th."

144, *l.* 18, *for* "1684," *read* "1648."

145, *note* ‡, *for* "vide Appendix," *read* "Ibid."

155, *l.* 28, *for* "Grindali," *read* "Grindalli."

158, *last line*, *for* "bordar," *read* "bordure."

175, *l.* 14, *for* "miro," *read* "miror."

183, *l.* 6, *for* "Comb," *read* "Coombe."

189, *l.* 14, *for* "Resurrecto," *read* "Resurrectio."

— *l.* 15, *for* "Rego," *read* "Regno."

— *l.* 17, *for* "Januari," *read* "Marij."

200. The error alluded to has since been corrected.

203, *l.* 14, *for* "who departed," *read* "and departed."

209. I find, on examining the inscription commemorative of John Redynge, Esq., given in this page, that it contains more than one error. It is well known that Henry VIII. had no other son than Edward his successor; therefore I would read Henry VII. for Henry VIII., and as Redynge is termed "late treasurer to *prince* Henry," it is clear that he did not live to see his master on the throne; consequently, the date of his death is also incorrect. Perhaps we should read 1508 for 1680, an error not unlikely to occur in transcription, and which is rendered more probable by the absence of the name of Redynge from the church register of the latter year.

288, *l.* 11, *for* "30 Eliz." *read* "38 Eliz."

The New Church at Croydon.

CHAPTER I.

Early History and present State.

Croydon, formerly called Croindene, Croiden, Crondon, and by the learned Camden, Cradiden, (Saxon Cꞃoᵹdæne), a large and handsome market town, lying nine miles and a half south of London, is situated on the edge of Bansted Downs, and contiguous to the spring-head of the river Wandle. The parish is about thirty-six miles in circumference, and covers a space of nearly ten thousand acres. It is bounded, on the east, by Beckenham and Wickham; on the west, by Mitcham and Beddington; on the north, by Streatham and Lambeth; and on the south, by Couldson, Addington, and Sandersted; and includes the hamlets of Addiscombe, Croham, Coombe, Haling, Shirley, Woodside, Waddon, Thornton Heath, Broad Green, and Barrack Town; the manors of Waddon, Whitehorse, Norbury, Haling, and Croham; and part of Norwood.

Nothing can be affirmed with certainty, in regard to the derivation of the name of this place: Ducarel was inclined to think that it might have its origin from the old Norman or French word Craye or Craire, chalk, and the Saxon word Dun, a hill; which supposition may be strengthened by the fact, that a large quantity of chalk has, till lately, been dug out here: whilst another writer derives its name from Crone, sheep, and Dene, a valley.

The locality of the ancient city of Noviomagus, mentioned in Antinonus' "Itinerary*," was supposed by Talbot* and Dr. Stukely† to have been at or near Croydon‡; whilst Somner, Burton, the

* Burton's Commentary on Antinonus' "Itinerary," p. 373.

† Dr. Stukely, in his "Itinerarium Curiosum," placed it at Crayford, but afterwards retracted his opinion. Vide Ducarel's Hist. of Croydon, p. 2.

‡ Several Roman pieces of money have been found here—two gold

Bishops Stillingfleet and Gibson have placed it at Crayford in Kent; and Camden, Gale, and Horsley, at Woodcote.

As the time when this city was known to exist is so far removed, its situation so unsatisfactorily described, and the opinions of these celebrated antiquaries so various, it is more than probable that its site may ever remain involved in its present obscurity.

The manor of Croydon was in the possession of the see of Canterbury, so early as the reign of William the Conqueror. For, in Domesday Book we read "In the hundred of Waleton*, Archbishop Lanfranc holds Croindene in demesne. In King Edward the Confessor's time, it was rated at 80 hides, and now at 16 hides and 1 virgate. The arable is 20 carucates. In demesne, are 4 carucates and 48 villans, and 25 bordars with 34 carucates. Here is a church and one mill, in value *5s.;* 8 acres of meadow and wood for 200 hogs. Of the land of this manor, Restoldus holds seven hides of the archbishop, and Radulphus one hide,

coins of Valentinian and a brass coin of Trajan, were dug up in the town in 1791, (see also "Whitehorse," Chap. III.); and Bray, in the "Archæologia, Vol. IX. p. 104, affirms, that the Roman road running from Arundel to London, and passing through this city, might be traced on Broad Green.

* Croydon is in the hundred of Wallington (Waleton), now a small hamlet in the parish of Beddington.

they have from thence 7*l.*, and 8*s.* of gabel (or tax). The whole, in the time of King Edward the Confessor, was worth 12*l.*; now 27*l.* to the archbishop, to his men, 10*l.* 10*s.*" In 1291, it was taxed at 20*l.**; in 1322, at the same†; in Archbishop Bourchier's time, it was valued at 55*l.* 3*s.* 11*d.* *per annum*‡; and in the Survey in 1646, at 274*l.* 19*s.* 9½*d.*, exclusive of the timber.

The parish is assessed at the sum of 1,444*l.* 9*s.*, to the land-tax, and is within the jurisdiction of the county magistrates, of whom those acting for the division hold a petty session weekly.

The town extends about a mile in length. The present High Street was originally only a bridle-way running through the fields. The old or lower town, called Old Croydon, formerly covering the same space, was situated farther from London than New Croydon §, and reached a great way towards Beddington, the ruins of which were standing in 1783||.

The houses are for the most part well built, the streets generally paved, the inns large and well conducted. The town, having lately been lighted with gas, presents altogether a very dif-

* Vide "Bodleian Valor," Bodleian Library.

† Reg. Reynolds, fol. 79. b.

‡ Cart. Misc. Lamb. MS. Library, Vol. XIII. No. 14.

§ Talbot, Leland's Itinerary, Vol. III. p. 176.

|| Ducarel.

ferent appearance to that ancient Croydon, described in an account written in the reign of Queen Elizabeth, which says: "The streets were deep hollow ways and very dirty, the houses generally with wooden steps into them—and the inhabitants in general were smiths and colliers."

It appears by the termination of this sentence, that the inhabitants were, in former times, almost exclusively employed as colliers or charcoal burners, a trade for which they have been celebrated by several of our elder poets; among others, by Alexander Barklay, in his Egloges, published about 1550*. Thomas Peend in his fable of "Hermaphroditus and Salmacis" published in 1565, says, that Vulcan

> "A Croydon Sangwine right did seme."

And Greene, in his "Quip for an Upstart Courtier," published in 1592, has the following—"Marry, quoth hee that lookt like Lucifer, though I am black, I am not the divell, but indeed a collyer of Croydon."

In the tragedy of "Locrine" published in 1595, and by some erroneously attributed to Shakespeare, we find the following distich:—

> "The colliers of Croydon,
> The rustics of Roydon."

* Vide page 23.

Besides which we have the following, in no way favourable, description of Croydon and its colliers, in a volume of poems, called "The Nightingale, Sheretine, and Mariana, &c." by one Patrick Hannay, published in 1622.

"In midst of these stands Croydon, cloth'd in blacke,
In a low bottome sinke of all these hills;
And is receipt of all the durtie wracke,
Which from their tops still in abundance trills,
The unpav'd lanes with muddie mire it fills
If one shower falls; or, if that blessing stay,
You may well smell, but never see your way.

"For never doth the flowre-perfumed aire,
Which steals choice sweets from other blessed fields,
With panting breast take any resting there,
Nor of that prey a portion to it yields;
For those harsh hills his comming either shields,
Or else his breath, infected with their kisses,
Cannot inrich it with his fragrant blisses.

"And those who there inhabit, suting well
With such a place, doe either Nigros seeme,
Or harbingers for Pluto, prince of hell;
Or his fire-beaters one might rightly deeme;
There sight would make a soule of hell to dreeme,
Besmear'd with sut, and breathing pitchie smoake,
Which (save themselves) a living wight would choke.

"These, with the demi-gods, still disagreeing
(As vice with virtue ever is at jarre)
With all who in the pleasant woods have being,
Doe undertake an everlasting warre,
Cut down their groves, and after doe them skarre
And in a close-pent fire their arbours burne,
Whileas the muses can doe nought but mourne.—

"The other sylvans, with their sight affrighted,
Doe flee the place whereas these elves resort,
Shunning the pleasures which them erst delighted,
When they behold these grooms of Pluto's court;
While they doe take their spoiles, and count it sport
To spoil these dainties that them so delighted,
And see them with their ugly shapes affrighted.

"To all proud dames, I wish no greater hell
Who doe disdaine of chastly profered love,
Then to that place confin'd there ever dwell;
That place their pride's deare price might justly prove,
For if (which God forbid) my dear should move
Me not come nie her,—for to passe my troth,—
Place her but there, and I shall keepe mine oath."

And again—there is a comedy intitled "Grim, the Collier of Croydon, or the Devil and his Dame, with the Devil and St. Dunstan, by J. T.," published in 1662.

When the inhabitants discontinued their sooty avocations, I have not been able to learn; but Ducarel, writing in 1783, says, Croydon "is surrounded with hills well covered with wood, whereof great store of charcoal is made."

Within this parish and manor are seven boroughs: Coombe, Selsdon, Bencham, Addiscombe, Woodside, Shirley, and Croham; from each of which a constable is annually appointed at the general court leet held for the manor of Croydon in Easter week, when a head constable, two petty constables, and two head-boroughs are nominated for the last-mentioned place.

There are also eight beadlewick lands, the owners of which, in their turn, serve the office of beadle; they collect the fines and amerciaments, but, with the reeves, receive no emolument.

The customs of the manor are as follows:—

1. One heriot, being the best beast of every copyholder dying seised of any messuage or tenement, not lying within the four crosses*, shall be paid for every such messuage or tenement; and if he have no quick cattle, then three shillings and sixpence for a dead heriot.

2. On the death of every copyholder for life, three shillings and sixpence for a dead heriot, and no more.

3. If any person to whom a right of copyhold shall descend shall die before admittance, one quick heriot is due for every messuage or tenement, and no more; and for want of a quick heriot, three shillings and sixpence for a dead heriot.

4. If a surrender be made of a copyhold to any person being no copyholder before, then, he is to

* These crosses were thus described in the reign of Elizabeth: the first is at Burchall's House, in an elm tree—the second is at the pound—the third is at Little Alms House Corner—the fourth at Dodd's corner, in an elm tree, against the Catharine Wheel Corner. Their exact situation cannot now be discovered; but the copyhold estates lying within the square originally formed by these four crosses, are exempt from the payment of heriots.

fine at the will of the lord, and to pay three shillings and sixpence for a dead heriot, and no relief.

5. If a surrender be made of a copyhold to any copyholder, there is due to the lord three shillings and sixpence for a dead heriot, and a relief, which is the extent of the rent (*i. e.* the quit-rent) by the year due to the lord, and no more.

6. Copyholds descend to the youngest son; and, no son, then to the youngest daughter; and so to the youngest in every degree.

7. All copyholders who have any estate of inheritance, may strip and waste, but the tenant for life may do neither.

8. No copyholder may let a lease of his copyhold, without licence of the lord, for more than three years, and is to give to the lord for every year that he is to have licence to let his copyhold, sixpence, and no more.

The quit-rents are collected by the reeves annually chosen by the homage jury, at the general court baron; there are eight reeveswick lands; the reeves are generally chosen in rotation.

There were anciently some grammar-schools in this town, for, in the register of Archbishop Courtney*, it is recorded, that on the 31st of May, 1393, John Makheyt, master of the grammar-schools at Croydon, was ordained a deacon

* Fol. 182 b.

at Maidstone—but of these, no traces are now to be discovered.

Besides the school founded and endowed by Archbishop Tenison*, there is one upon the Lancastrian system, established in 1812, for the education of indigent children of every religious persuasion; for which institution, a school-house was erected at North End in 1829. Another, the national or parish charity school, conducted on the principle of the late Dr. Bell, and instituted in 1812, is now held in the school-house adjoining and belonging to Archbishop Whitgift's Hospital. In addition to these, there is a school of industry for female children conducted in the palace chapel, and an infants' school under the patronage of the ladies—all which are supported by voluntary contributions.

The Society of Friends have also a large establishment, situated in Park-lane; which was removed here in 1825, from Islington, where it had existed for more than a century. It is supported by subscription, and provides for the education and maintenance of 150 boys and girls.

On the common has lately been erected, partly from a grant of 3,500*l.*, and partly from a loan to be paid off by instalments from the parliamentary commission, a chapel of ease, dedicated to St. James, after a design of R. Wallace, Esq.

* Vide post, Chap. IV.

The first stone was laid by the Rev. J. C. Lockwood, 16th May, 1827, and the consecration performed by his present Grace, 31st Jan. 1829. It is a brick building, of the pointed style of architecture, and has a small campanile tower at the west end, with pinnacles at each angle. It contains 1,200 sittings, of which 400 are free.

A chapel of the same order, dedicated to St. Peter, was erected about the same time on Beulah-hill, after a design of J. Savage, Esq., from funds provided in a similar manner. The first stone was laid 12th Nov. 1827. This elegant edifice contains 1,500 sittings. It has a small tower at each extremity, and several richly crocketed pinnacles at the west front. These two chapels are perpetual curacies, in the patronage of the vicar of Croydon.

Several denominations of dissenters have meeting-houses here: the Quakers, in Park-lane, which is numerously attended; the Independents have a neat Gothic chapel in George-street; the Wesleyan Methodists, a chapel at North End, erected in 1829; and the Anabaptists, at Pump Pail. There are some others of minor importance.

The barracks, erected in 1794, are at the entrance of the town by Mitcham, and were originally intended as a temporary station for cavalry, during the preparation of troops for foreign service. They contain accommodation for three troops of cavalry, with a hospital for 34 patients, an infirmary, stabling for 192 horses, a store-room for 1,000 sets of harness with field equipments, riding-house, and the accustomed offices.

The present court-house and corn-market, built after a design of the late Samuel Pepys Cockerell, Esq., at an expense of about 8,000*l.*, defrayed by the proceeds accruing from the sale of waste land belonging to the parish, disposed of by act of Parliament in 1806, was first opened in the summer of 1809. It is a neat stone edifice surmounted by a cupola containing a clock, and comprises, in the upper story, a court for the trial of civil causes at the county assizes (which are held here alternately with Guilford), with rooms for the judges, sheriffs, and grand jury; where is also held, every alternate week, the Court of Requests for the recovery of debts under 5*l.* The ground-floor is reserved for a corn-market, and during the assizes is appropriated to the Criminal Court.

The old market house, built at the cost of Francis Tirrell, citizen, in the year 1566, was pulled down in 1807, when the following inscription was discovered—

"This Markett House was buylt att the coste and charges of Francis Tirrell, citizen and grocer of London, who was born in this towne and departed this worlde in Sept. 1609."

The butter market, situate in High Street,

was erected in 1808, at a cost of 1,219*l*, the money arising from the same fund. The markets are held on Saturday.

The prison, situated behind the corn market, and occupying the site of the Old Town Hall, is a substantial brick building, erected in 1803 by a subscription among the inhabitants, the lower part only of which is used for the confinement of prisoners, the upper being let as a warehouse.

The little alms houses, described, in 1722, as "nine small low inconvenient houses, wherein are usually placed the parish poor," are situated near the church, and must have been built previous to 1528, as in that year a rent-charge of 1*l.* was given to this charity by Joan Price. In 1629, Arnold Goldwell gave 40*l.* towards their re-erection; and, in 1775, they were enlarged by the addition of two new buildings for twelve poor residents, with funds received from the then Earl of Bristol, and a voluntary subscription among the principal inhabitants.

The workhouse, situated on Duppas Hill, to the westward of the town, was erected about 1727, on a piece of ground given (*inter alia*) by deed in 1629, by Sir William Walter, to the inhabitants of Croydon, for the purpose of digging gravel for mending the parish roads and other uses, and is capable of accommodating above 160 persons.

There is an iron rail-road passing from Wandsworth through this town, to the chalk-pits at Merstham; also a canal, opened 22nd October, 1809, which, after running from the north end of the town through Norwood, Penge Common, Sydenham, Forest Wood, and New Cross, falls into the Thames at Rotherhithe.

The population of Croydon, as shewn by the census of 1831, amounts to 12,447* inhabitants, and 2,431 houses. In 1783, there were between 700 and 800 houses in the town, and the inhabitants were computed at rather more than 5 to each † house. In 1801, the census returned 5,743 inhabitants, and 1,074 houses; in 1811, 7,801 inhabitants, and 1,474 houses; and in 1821, 9,254 inhabitants, and 1,639 houses.

By the recent Reform Act, 2 Will. IV. c. 45, Croydon is appointed one of the polling places for the eastern division of the county of Surrey ‡.

We have been told §, but I know not on what authority, that King James I., the first institutor of regulations relating to horse-racing, held Croy-

* An extra-parochial spot, between Croydon and Addington, contains about 100 inhabitants.

† Ducarel's Hist. of Croydon, Appendix, p. 156.

‡ The Members returned were—John Ivatt Briscoe, Esq., and Aubrey William Beauclerk, Esq.; the numbers at the conclusion of the poll being, for Briscoe, 1667; Beauclerk, 1163; Allen, 849; and Lainson, 250. The votes of the parish of Croydon were as follows—for Briscoe, 174; Beauclerk, 117; Allen, 67; Lainson, 15.

§ Vide "The Horse," published by the Society for the Diffusion of Useful Knowledge.

don and Enfield chase in the greatest estimation as resorts for this his favourite pastime. The amusements of Croydon are now confined to occasional assemblies held at the Greyhound; the theatre, erected in 1800, which has of late years been but seldom open; and the fairs, one, the principal, held on the 2nd of October, and continuing for three days, and the other on the 5th of July and two following days.

During the great Rebellion, General Fairfax, for a time, occupied Croydon as his head quarters; whence he marched August 10, 1647, to Kingston, where he held a council of war the next day*.

* Vide Perfect Occurrences, August 6—13, 1647.

CHAPTER II.

Chronology.

Having given a brief sketch of the town and some of its public buildings, I shall reserve for a future chapter a detailed description of the church, palace, and charitable institutions; and proceed to lay before the reader a chronological account of the remarkable events relating to this populous town.

1185—The town amerced one mark for a default*.

1200—Two women, having stolen some clothes at Croindone, were pursued to Southfleet, where they were seized, imprisoned, and afterwards tried by the Lord Henry de Cobham and other gentlemen of the county, who adjudged them to undergo the fire-ordeal (*ad portandum calidum ferrum*). By this cruel and superstitious test of innocence, one was acquitted, the

* Vide Madox's History of the Exchequer, p. 384.

other condemned and afterwards drowned in a pond called Bikepool*.

1264—On the 14th of May, the Londoners, flying from the battle of Lewes, where they had taken part with the barons against Henry III., were intercepted at Croydon by a detachment of the king's forces then lying at Tonbridge castle, who put them to the sword with great slaughter†.

1270—In this year, John, the celebrated and seventh Earl Warren, dated an instrument from *Creyndone*‡, stating his intention to stand to the judgment of the court, after his outrage on Alan Lord Zouch of Ashby and his son Roger, on pain of excommunication and forfeiture of his estates§.

1273—Archbishop Kilwardby obtained a grant of a market to be held every Wednesday ||.

1276—This same archbishop obtained a grant

* Vide Blount's "Ancient Tenures and Customs of Manors."

† Vide Holingshed's Chronicles, ed. 1585, Vol. III. fol. 269.

‡ Claus. 54 Hen. III. m. 5.

§ This outrage, committed in Westminster-hall, was occasioned by an estate being adjudged to Lord Zouch, which was unjustly claimed by the fiery earl, who, when the verdict was pronounced, giving loose to the natural vehemence of his temper, drew upon that nobleman and his son, and almost killed the father, and severely wounded the son. He was fined 10,000 marks, which the King afterwards remitted to 8,400.

|| Cart. 5 E. I. m. 24.

of a fair to be held for nine days, beginning on the 16th of May*.

1286—On the 15th of December, William, only son of John Plantagenet, seventh Earl of Warren and Surrey, was unfortunately killed in a tournament at Croydon. Stowe has thus recorded his death:—"An. reg. 15 Ed. I. William Warren, sonne and heyre of John Warren, Earl of Surrey, in a tornement at Croyden, was by the challenger intercepted and cruelly slaine†."

1314—Archbishop Reynolds obtained a grant of a market to be held here on Thursday, and a fair on the vigil and morrow of St. Matthew's Day‡.

1343—Archbishop Stratford obtained a like grant of a market to be held on Saturday, and a fair on the feast of St. John the Baptist§.

1352—On the 18th of February, Archbishop Islip granted to Robert Farnham and William Chober, for the term of their natural lives, a messuage and nine acres of land in Croydon, which had escheated to him upon the death of John Latyn, Silvestria his wife, and their son

* Cart. 5 E. I. m. 24.

† Stowe's Annals, p. 311; Watson's Lives of the Earls of Warren and Surrey.

‡ Cart. 8 E. II. m. 15.

§ Ryley, p. 586.

William, to whom a like grant of the premises had been formerly made by Archbishop Stratford*.

1362—On the 22nd of February, Archbishop Islip granted to Thomas de Kendale a messuage and nine acres of land, with their appurtenances, in Croydon, which escheated to him on the death of the before-mentioned John Latyn, Silvestria his wife, and William their son, for a hundred years, paying ten shillings rent *per annum*†.

1382—On the 20th of March, Sir William Walworth, the patriotic Lord Mayor of London, was appointed Keeper of Croydon Park‡.

1412—On the 30th of November, the unfortunate James I. of Scotland, signed at this place a deed of general confirmation to Sir William Douglas of Drumlanrig; by which it is evident that he was then at Croydon palace in the custody of Archbishop Arundel. This grant, preserved in facsimile in Anderson's "*Diplomatum et Numismatum Scotiæ Thesaurus*" is as follows:—

"Jamis, throu the grace of God Kynge of Scottis, Till all that this lettre heris or seis sendis gretynge. Wit ze that we have grauntit, and be this presentis lettres grauntis a speciall confirmatin in the maiste forme till oure traiste and wele belofit cosyng Sir William of Douglas of Drumlanrig of all the landis that he is possessit and

* Reg. Islip. fol. 44 a. † Cart. Miscell. Vol. X. No. 20.
‡ Reg. Courtney, fol. 37 a.

[illegible]

[illegible].—John Hewes, a [illegible] of London, was made to abjure, for saying that he heard the vicar of Croydon [Phillips] preach openly, "that there is as much idolatry kept by going in pilgrimage to Wilsdon or Muswell as in the stews beside, &c."*

About this time, at the palace, John Frith, who afterwards suffered martyrdom at Smithfield, July 4, [illegible], was heard in defence of his opinions, before the commissioners, Archbishop Cranmer, Lord Cromwell, Stokesley bishop of London, Dr. Heath, and others, appointed by the king for that purpose. When travelling here from the Tower in custody of two of the archbishop's household, they, fearing for his life, advised him to make his escape: but he refused, telling them "that if they went away and left him alone, he would come to Croydon himself, and appear before the bishop†."—On the

* Fox, Vol. II. p. 592.

† "Epitomy of Ecclesiastical History." By J. Shurley. 1683. Part 2, p. 39.

night of his arrival at Croydon, as we are told by Fox, who gives a very minute account of this proceeding, "he was well entertained in the porter's lodge*."

1542—On Trinity Sunday, Archbishop Cranmer had before him, at his consistory at Croydon, all the prebendaries and preachers of Canterbury cathedral, when he argued with them concerning the diversity of their doctrines†. Their names were—Richard Thornden, Arthur Sentleger, Richard Parkhurst, Nicholas Ridley, John Meines, Hugh Glazier, William Hunt, William Gardiner, John Milles, John Daniel, Robert Goldson, John Baptist—prebends; Robert Serles, Michael Drune, Lancelot Ridley, John Scory, Edmund Shether, and Thomas Brooke—preachers.

1545—On the 16th of October, a commission of array was issued for raising 400 able men, when this town was required to furnish four archers and six billmen.

1551—On the 25th of May, Croydon and its neighbouring villages experienced a shock from an earthquake‡.

1552—On the 10th of June, Alexander Barclay, or De Barklay, D.D., author of "The Gret

* Fox. Vol. III. p. 1927. † Strype's Life of Cranmer, p. 108.
‡ Bishop of Hereford's Annals.

Shyppe of Fooles of this Worlde," "Myrrour of Good Maners*," &c., was buried here†. This elegant writer is supposed to have been a native of Scotland. In 1495, he entered himself of Oriel College, Oxford, and having distinguished himself by the quickness of his parts, he quitted England for the continent. Upon his return home, he was appointed chaplain to the bishop of Wells, who made him one of the priests of St. Mary at Ottery in Devonshire. He was afterwards a monk of the order of St. Benedict—a Franciscan, and finally a monk of Ely, upon the dissolution of which monastery, in 1539, he was collated to the vicarage of St. Matthew, at Wokey, Somerset. In February, 1546, being then D.D., he was presented to the vicarage of Much-Badew or Baddow-Magna, Essex; and, on the 30th April, 1552, to the rectory of All-hallows, Lombard-street, which he did not enjoy above the space of

* The other works of this writer are—The famous Cronycle of the Warre which the Romans had agaynst Jugurth, usurper of the kyngdom of Numidy: which cronycle is compyled in Latyn, by the renowned Romayn Sallust—Orationes Variæ—De Fide Orthodoxa—The Castell of Laboure, wherein is rychesse, virtue, and honour:—The Figure of our Mother Holy Church oppressed by the French King—Answer to John Skelton, the Poet—The Lives of S. Catherine, S. Margaret, and St. Etheldred—The Life of S. George.—De Pronuntiatione Gallica.—The Miseries, or Miserable Lives of Courtiers.

† Vide Parish Register. Wood (Athen. Oxon. Vol. I., p. 207) says he was buried *in* the church.

six weeks, dying in this town in the month of June the same year. In his Egloges, we find the following separate lines relative to this place; in one of which, he informs us of his having resided here in early life—

> "And as in Croidon I heard the collier preache."

> "While I in youth in Croidon towne did dwell."

> "He hath no felowe betwene this and Croidon
> Save the proude plowman Gnatho of Chorlington."

1567—Queen Elizabeth visited Archbishop Parker at Croydon, where, on the 30th of April, she held a council*.

1573—On the 4th of July, Queen Elizabeth, with all her retinue, came to Croydon palace on a visit to the same archbishop; with whom she remained seven days previous to her going a progress into Kent†. From the following MS. presented by the Reverend Dr. Birch to Archbishop Herring, and now in the library of Lambeth palace, it appears that her majesty intended to favour the people of Croydon with her presence the year ensuing—if she did not actually do so.

* Council-book in the Duke of Buckingham's library at Stowe.

† Parker's Antiquit. Brit. ed. Drake, pp. 553, 554.

"Lodgins at Croyden, the Busshope of Canterburye's house, bestowed as followeth, the 19th of Maye, 1574—

The Lord Chamberlayne* his old lodginge
The L Tresurer† wher he was
The La Marques‡ at ye nether end of the great chamber
The La of Warwick wher she was
The Erle of Lecester§ wher he was
The Lord Admyrall|| at ye nether end of ye great chamber
The La Howard wher she was
The Lo of Honsdane wher he was
Mr. Secretarye Walsingham wher Mr. Smyth¶ was
The La Stafforde wher she was
Mr. Henedge** wher he was
Mrs. Drewreye wher ye La Sydney was
Ladis and gentylwomen of ye Privye Chamber ther olde
Mrs. Abbington her olde, and one other small rome added for ye table
The maydes of honnor wher they wer
Sir George Howard wher he was
The Capten of ye gard†† wher my L of Oxford was
The gromes of ye Privye Chamber ther olde
The esquyeres for the bodye ther olde
The gentylmen husshers ther olde
The phesycyas ij chambers

---

* Thomas Radclyffe, Earl of Sussex.

† William Cecil, Lord Burleigh.

‡ Elizabeth Paulet, Marchioness of Winchester.

§ Robert Dudley.

|| Edward Fynes, Earl of Lincoln.

¶ Afterwards Sir Thomas, and Secretary of State.

** Afterwards Sir Thomas, Vice Chamberlain.

†† Christopher Hatton, Esq., afterwards Sir Christopher.

The Quen's robes wher they were
The grome porter wher he was
The clark of the Kytchen wher he was
The wardrobe of bedes.

For the Quen's Wayghters, I cannot as yet fynde anye convenyent romes to place them in, but I will doo the best y[t] I can to place them elsewher, but yf y[t] please you S[r] y[t] I doo remove them. The Gromes of the Privye Chamber nor Mr. Drewrye have no other waye to ther chambers but to pas thorowe that waye agayne that my Lady of Oxford should come. I cannot then tell wher to place Mr. Hatton; and for my La Carewe here is no place with a chymeney for her, but that she must ley abrode by Mrs. Aparry and the rest of y[e] Pryvy Chambers. For Mrs. Shelton herę is no romes with chymeneys; I shall staye one chamber without for her. Here is as mutche as I have any wayes able to doo in this house. From Croyden this present Wensday mornyinge, your Honnors alwayes most bowden S. Bowyer."

1577—On the 17th of April, Sir Richard Gurney, the celebrated Lord Mayor of London, was born in this town*. This patriotic citizen was, for his obstinate devotion to royalty, deprived of his mayoralty, rendered incapable of holding any public office in the kingdom, fined 5,000*l.*, and imprisoned during the pleasure of both houses of Parliament in the Tower, where he remained till within a month of his death, a term of seven years. Sir Richard, who was created a baronet by Charles I., died October 6th, 1647, having suffered in his estate to the extent

* Lloyd's Memoirs, pp. 625, 626.

CHAPTER II.

Chronology.

Having given a brief sketch of the town and some of its public buildings, I shall reserve for a future chapter a detailed description of the church, palace, and charitable institutions; and proceed to lay before the reader a chronological account of the remarkable events relating to this populous town.

1185—The town amerced one mark for a default*.

1200—Two women, having stolen some clothes at Croindone, were pursued to Southfleet, where they were seized, imprisoned, and afterwards tried by the Lord Henry de Cobham and other gentlemen of the county, who adjudged them to undergo the fire-ordeal (*ad portandum calidum ferrum*). By this cruel and superstitious test of innocence, one was acquitted, the

* Vide Madox's History of the Exchequer, p. 384.

other condemned and afterwards drowned in a pond called Bikepool*.

1264—On the 14th of May, the Londoners, flying from the battle of Lewes, where they had taken part with the barons against Henry III., were intercepted at Croydon by a detachment of the king's forces then lying at Tonbridge castle, who put them to the sword with great slaughter†.

1270—In this year, John, the celebrated and seventh Earl Warren, dated an instrument from *Creyndone*‡, stating his intention to stand to the judgment of the court, after his outrage on Alan Lord Zouch of Ashby and his son Roger, on pain of excommunication and forfeiture of his estates§.

1273—Archbishop Kilwardby obtained a grant of a market to be held every Wednesday ||.

1276—This same archbishop obtained a grant

* Vide Blount's "Ancient Tenures and Customs of Manors."

† Vide Holingshed's Chronicles, ed. 1585, Vol. III. fol. 269.

‡ Claus. 54 Hen. III. m. 5.

§ This outrage, committed in Westminster-hall, was occasioned by an estate being adjudged to Lord Zouch, which was unjustly claimed by the fiery earl, who, when the verdict was pronounced, giving loose to the natural vehemence of his temper, drew upon that nobleman and his son, and almost killed the father, and severely wounded the son. He was fined 10,000 marks, which the King afterwards remitted to 8,400.

|| Cart. 5 E. I. m. 24.

to the vicarage of St. Martin in the Fields, being at that time D.D.; he was afterwards ordained archdeacon of Lewes; and on the 10th December, 1660, dean of Rochester. He was also rector of Henley, Oxon. During the commonwealth, he officiated as minister of St. Dionyse, Back-church, London. "At length," says the author above quoted, "this active and forward man, who had little or no character among the true loyalists, especially that part of the clergy who had suffered in the times of usurpation, giving way to fate in his house at Croydon in Surrey, on the first day of June, in sixteen hundred and seventy, was buried on the 9th day of the same month in the chancel of St. Martin's church in the Fields."

1687—On the 29th of April, Sir Christopher Hatton was appointed Lord Chancellor "at Croydon in the Archbishop of Canterburie's house, where he received the great seale in the gallery there*."

1728—On the 12th of May, so violent a storm of hail and rain, with thunder and lightning, fell at Croydon, as to strike the hail-stones, which were from eight to ten inches round, some inches

* Stowe's Annals, p. 742.

into the earth. The cattle were forced into ditches and drowned, the glass windows facing the storm were shattered, and other great damage done.

1744—Much damage was done by lightning in and near Croydon.

CHAPTER III.

Manors and Park.

On the inclosure in 1797, claims were made and allowed for the following manors:—

1 Croydon*, by the Archbishop of Canterbury†.
2 The Rectory‡, by Robert Harris, Esq.
3 Waddon, by the Archbishop of Canterbury.
4 Whitehorse, by John Cator, Esq.
5 Norbury, by Richard Carew, Esq.
6 Haling, by William Parker Hammond, Esq.
7 Croham, by the Warden and Poor of Whitgift's Hospital.

Waddon,

Anciently styled Woddens, is a considerable manor lying about half a mile from the town on the road to Beddington, and contains several gentlemen's seats, a large water-mill men-

* For an account of this manor and its customs, vide ante, p. 3.

† Dr. John Moore.

‡ An account of this manor will be found in Chapter VII.

tioncd in Domesday book and belonging to the archbishop, and many respectable houses. In 1127, this manor was given by Henry I. to the monks of Bermondsey*. "Whether," says Salmon, "this was the whole manor is a question, because there appears a confirming charter of Henry II., of half the manor of Wedone, the gift of Roger de Thebovill to the Abbey of Becc† " This, however, may be Whaddon, of which name there are manors in various counties, more especially as the Convent of St. Saviour conveyed the whole manor in 1390, to Archbishop Courtney, in exchange for the appropriation of the church of Croydon; from which time it has been annexed to the see of Canterbury. In the time of Henry IV. this manor was taxed at 10*l.* 16*s.* 5*d.* In Archbishop Bourchier's time, it was valued at 8*l.* 12*s.*‡; and in that of Archbishop Parker, at 22*l.* 6*s.* 8*d.*§

A court baron is annually held here in Easter week.

* Mon. Ang. Vol. I. p. 639, 642.

† Antiquities of Surrey, p. 42.

‡ Cart. Miscell. Lamb. MS. Lib. b. 13, No. 14.

§ Lamb. MS. Lib. No. 1142.

WHITEHORSE.

This manor, also called Bunchesham, and Bencham, is situate about a mile and a half north from the town, on the road to Norwood. The mansion appears, from the date 1604, formed in the brick-work of the south gable, to have been built early in the reign of James I. In 1287, Jeffry de Haspale held the manor in fee, to be inherited by William and Phillip de Padyndenne*, which latter held it at his death in 1309†. Yet Peter Chaceport had a grant of free warren here in the reign of Henry III.‡, as had Richard de Gravesend, Bishop of London, in the reign of

* Esch. 2 Ed. I. n. 25. † Esch. 2 Ed. II. n. 50.
‡ Cart. 37 Hen. III. m. 15.

Edward I.*, Stephen de Gravesend, bishop of London, died in the reign of Edward III., holding this manor of the manor of Croydon, for the service of twenty-one shillings *per annum*, with suit of court to the Archbishop at Croydon, from three weeks to three weeks. There was then a capital messuage of no value beyond reprises; 200 acres of arable land worth 58*s*. 4*d*. *per annum*, of which 100 was valued at 4*d*., the other 3*d*. an acre; the pasture of 8 acres of wood 12*d*.; the pannage when it happens, *communibus annis*, 18*d*.; the underwood, 4*s*.; 8 acres of meadow land, 8*s* ; 20 acres of pasture, 3*s*. 4*d*.; rents of assize as well from free tenants as from natives, 70*s*.; at Christmas, 24 hens and 1 cock, 4*s*.; at the same time, 6 ploughshares, 4*s*.; pleas and perquisites of Courts, 3*s*. 4*d*. It was then found that the reversion belonged to Hugh de Nevill by fine levied in the King's court †. It was next in the possession of the Cherburys ‡, afterwards of the Chiritons §, one of whom alienated it to Walter Whitehorse, the king's shield-bearer, who also obtained a grant of free warren ||. Arnold Holker possessed the manor in the reign of Henry IV., and had a confirmation in fee of free warren ¶;

* Cart. 27 Ed. I. n. 6.
† Esch. 12 Ed. III. n. 34.
‡ Cart. 29 Ed. III. m. 9.
§ Cl. 41 Ed. III. m. 6.
|| Pat 43 Ed. III.
¶ Pat. 7 Hen. IV. pt. 2, m. 36.

and in the reign of Henry VI., it was the property of Edmund Brudenell, who had a further confirmation of free warren*. In the 6 Henry VIII., Sir Robert Morton, Knt., nephew of Richard Morton, Bishop of Worcester, and grand nephew of Cardinal Morton, died seised of this manor*; and in the 9th year of the same reign, it was in the possession of John Morton, Esq., his half brother; from whom it descended to his nephew William; from him to his son William, who held it in 1566‡, and whose grandson Thomas, dying in 1678, left five daughters, amongst whom this estate was divided. Four of the five partitions were purchased by John Barrett, Esq., in 1712; the fifth, by his grandson in 1787, who sold the whole to John Cator, Esq., M.P.; whose nephew, John Cator, Esq., sold it to John Davidson Smith, Esq., the present possessor.

In 1719, a gold coin of the Emperor Domitian was found in this manor, and lately coins of Lælius Cæsar, and Titus Vespasian, with several others, all in good preservation.

No courts are held for this manor.

On this estate, a saline spring, long resorted to by the poor of its immediate neighbourhood on account of its medicinal properties, has lately been

* Pat. 10 Hen. VI. pt. 2, m. 6.

† Cole's Esch. Harl. MS. No. 576.

‡ Herald's Visitation of Surrey, 1623, Harl. MS. No. 1397.

brought into public notice as the "Beulah Spa." To its spirited proprietor, the inhabitants of the metropolis and its southern environs are greatly indebted. The grounds, which extend over twenty-five acres, and are entered by an elegant lodge, have been tastefully laid out under the direction of that accomplished architect, Decimus Burton. The picturesque character of the place, its rustic edifices, its rides, and its promenades, render it a most pleasing resort, not only for the invalid, but also for those who seek recreation and amusement.

CROHAM.

This manor, called also Cronham and Cranham, consists of a messuage and farm; it extends over Cromehurst, and is situate about a mile south east of the town; it receives quit-rents from several houses and lands in the town of Croydon, and forms part of the endowment of the hospital of the Holy Trinity. In 1368, one Chiriton alienated the manor to Walter Whitehorse*, the king's shield-bearer. It appears, however, to have reverted to its former possessor, as Edward III. in the 46th year of his reign, seized the manor, with other lands

* Ch. 41 Ed. III. m. 6.

then belonging to Walter Chiriton, for a debt of 3,000*l*. due to the crown, and granted it to John de Wesenham; but Richard II. restored it to William, son of the above Walter Chiriton. In the reign of Henry IV. it was again in the possession of the crown, when William Oliver was appointed keeper of the same*. By the court rolls of the manor, in the time of Henry VII., it appears to have been then the property of Lady Peche; and in the reign of Henry VIII., it was held by Sir John Danett, Knt., in right of his wife Anne, daughter and heir of Thomas Elmerugge, otherwise Elynbrugge, Esq., gentleman porter to Cardinal Morton. It afterwards became the property of Sir Olliphe Leigh, Knt., of Addington; who sold it to Archbishop Whitgift.

Courts are sometimes held for this manor, which is partly in the parish of Sandersted.

NORBURY.

The manor of Norbury, called also Northborough, lies on the west side of the London road, and extends over that side of Thornton Heath. Several houses and lands in Croydon pay quit-rents to it. In the 48 Edward III., Nicholas Carew† of Beddington, keeper of the privy

* Fin. 1 Hen. IV. m. 8.

† Cart. 48 Ed. III. n. 10.

seal, obtained a grant of free-warren on all his lands at Croydon. He died 17th August, 1391, seised of this manor, then consisting of a capital messuage, 100 acres of arable land, 300 acres of pasture, 10 acres of meadow, and 20 acres of great wood, lying in common, rents of assize 30*s*. He left Nicholas, his son and heir*, who died seised of the manor 33 Henry VI.†; whose son Nicholas also died seised of it, 6 Edward IV‡. On the execution and attainder of Sir Nicholas Carew, K.G., Henry VIII. seized it for his own use, and annexed it to his newly created honor of Hampton Court; and Edward VI., in the first year of his reign, granted it, with the meadow called Pyrlemead in Croydon, to the Archbishop of Canterbury, in performance of an agreement of his late father. In this same year, Sir Francis Carew, Knt. obtained a reversal of his father's attainder; but it does not appear that he was in possession of this manor till the reign of Mary; for, in the 6 Edward VI., 18th July, that king granted it, with the forfeited estate of Sir Nicholas, to Thomas, Lord Darcie, of Chiche, in exchange for other lands§. Queen Mary, in the second year of her reign, 14th July, (having, on the 20th No-

* Esch. 14 R. II. n. 10.

† Esch. 6 Ed. VI. n. 22.

‡ Esch. 6 Ed. VI. n. 40.

§ Pat. 6 Ed. VI. p. 9.

vember, the preceding year, obtained a reconveyance from Lord Darcie of the forfeited lands of Sir Nicholas), re-granted to Sir Francis Carew his father's estates; when Lord Darcie alienated this manor to him, 2nd January, 1556. From Sir Francis it has descended, with the Beddington estate, to its present possessor—Admiral Sir Benjamin Hallowell Carew, G.C.B., &c.

Haling*.

The manor of Haling is situate at the extremity of the town, and comprises a park and mansion. Towards the close of the fifteenth century, it was in the possession of Thomas Warham, Esq., who was one of the twelve principal inhabitants presenting to the chauntry of St. Mary in 1458†, and again in 1476‡. In his will, dated 3rd

* "*Inge* in nominibus locorum designat pratum, à Sax. inᵹ, (Vide Regulæ Generales, de Nominibus Locorum, ad finem Chron. Sax.) Sanctus in Saxon is haliᵹ, and from thence is derived the Old English word *All Hallows*, for *All Saints*, and therefore it is not unlikely that *halig* may mean the Holy Meadow, especially as it is not very far from a place called Woddens, in the map of Surrey (lately published by Bowen), which might induce a conjecture that here formerly was some idol of Woden (whence our Wednesday) adored in that place by the Old Pagan Saxons." Ducarel, p. 73.

† Reg. Bourchier, fol. 74 a.

‡ Ibid. 113 b.

September, 1478, and preserved in the chapter-house at Westminster, he directs his body to be buried in the parish church of St. John the Baptist of Croydon, in the chapel of St. Nicholas, before the image of our Lady of Pite. He gives legacies for masses, &c., with a distribution of torches used at his month's mind, amongst different churches. He also gives in lead, for covering the north aisle of the church of Croydon, four marks. By other papers, also in the chapter-house, we find that he held the manor of the archbishop by the rent of 21*s*. 0½*d*.; that the free rents and quit-rents paid to the said manor amounted to 12*s*. 8*d*.; that the clear yearly value amounts to 35*l*. 16*s*. 10½*d*.;—the house not accounted for—and that he had woods there, which, within ten years, would be worth 400 marks *per annum*.

"It is likely" says Bray*, "that he was father of William Warham, of whose lands of the manors of Halyng and Selerste, and of his lands in the towns of Croydon, Whaddon, and Mycham, there is an account amongst the above papers:" speaking of whom, he further says, "This William is probably the same as was appointed archdeacon of Canterbury in 1504, and held other valuable

* Manning & Bray's Surrey, Vol. II. p. 512.

church-preferments; all which he resigned on the death of Archbishop Warham." The pedigree of the Warham family runs thus:—

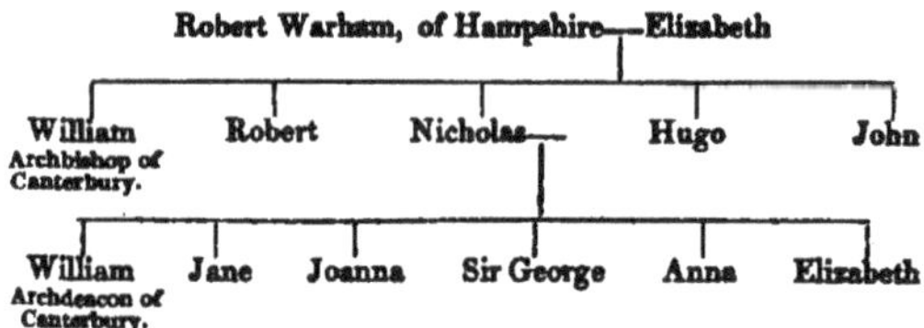

In these papers, it is said, that the manor-place of Halyng, with two orchards, two gardens, a culver-house with the bank of conies, were let to Sir Nicholas Carew, for 40*s.* *per annum;* the land of the said manor and game of conies (a high ridge of gravel soil, consisting of about twelve acres) at 12*l.*; and the farm of Selhurst, 12*l.* How Archdeacon Warham should become possessed of this manor, I am unable to tell, as the same was given by his uncle, the archbishop, to Henry VIII. in exchange for other lands*; perhaps we should read Archbishop for Archdeacon William Warham, and Thomas Warham as *uncle* of that prelate.

Queen Mary granted the manor of Haling to

* Grants and Exchanges of Lands, Aug. Office.

Sir John Gage, K.G.*, of whom his son, Robert Gage, Esq., of Haling, has left the following MS. account:—

"Sir John Gage of Furle, his Preferment at Court.

"Sir John Gage, Knight, my good father, whose soul God pardon, was, after my grandfather's death, warde to the duke of Buckingham; who, after my father was married to my mother, daughter to Sir Richard Guilfourd, Knight, was preferred by him to King Henry the Eighth his service; and after, he being at the wininge of Turwin and Turrein, was first made captain of the Castle of Callis; after he was made deputy of the Castle of Owns under my lorde Vawse. Shortly after he was sent for home, and presently made knight, of the privy counsell, vize chambelaine, and captain of the guard; within few yeares after, for service he did in the borders of Scotland, at his returne home was made controwler and chancellour of the Dowchye in one day; within few dayes after he was made counstable of the Tower of London; and the next St. George's feast after, knight of the most noble order of the garter. On goinge to wininge of Bullen he was joyned in the commission with Charles duke of Suffolke, lorde leauetenant of the king's majesty's campe, for sundry services there; with Sir Anthony Browne, knight, master of the horse and generall captayne of the bands of horsemen. After the death of our Soveraign lord king Edward the Sixth, at the cominge in of queen Mary, he was made her lord chamberlaine. Thus haveing served in all these roomes and offices truely and paynefully from the first yeare of the reigne of our soveraign lord King Henry the Eighth of famous memory, unto the fifth yeare of the reign of our soveraign lady queen Mary, untouched with any reproch or unfaithfull service in this time, being 77 yeares old, he ended his life, in favour of his prince, in his owne house at Furle in Sussex, committing his soul

To this memoir is appended the following memorandum:—

"This note was written by Robert Gage, of Halinge, in the county of Surry, esq., third son of the aforesaid Sir John Gage Knight, and Phillippa Guillfourd his wife; as John Gage of Halinge, eldest sonne of the said Robert Gage, my good father, hath divers times told me. In witness whereof, I under-written, son of John Gage aforesaid of Haleinge, and grand-child of the said Robert Gage, have subscribed my name, this 29th of January, 1630.

HENRY GAGE."

He left issue, four sons—Sir Edward, James, Robert, of Haling, M.P. for Lewes in 1533, and William. Robert died seised of this manor in 1587. He was father of John and Robert Gage, who were probably born here, the former of whom succeeded to the estate, and was father of colonel Sir Henry Gage, Knt., sometime governor of Oxford during the civil war, who was killed at Cullum bridge, near Abingdon, January 7th, 1644. The latter, during his travels abroad, having suffered imprisonment at Brussels for attempting the life of one Thomas Morgan, in the church of St. Guldula*, returned to England, and entering into Babington's conspiracy, was executed at St. Giles's in the Fields, 20th September, 1586. It does not appear, however, that he actually en-

* State of the English fugitives under the King of Spain, 1596.

tered into the conspiracy, but rather that he suffered as an accessory after the fact, in concealing the conspirators when their treasonable design had been discovered. In a MS. account* of their several trials and confessions, we read "that when all the matter was discovered, he lent Savage (who suffered for the same cause) a horse to flye to Croiden, and directed him to one off Savage's father's men, who should help him away." Among the charges urged against him at the trial, was, that he attended Ballard as his man when he went into the north to provoke the people to rebellion. He was discovered hid in a barn in Carnock's apparel, having lent his own to Babington. When asked on his trial wherefore he fled into the woods, he "stoutly and fiercely answered—For company†."

By the attainder of the above John Gage, Esq., the manor again reverted to the crown, and was leased to Charles, Earl of Nottingham‡, the celebrated lord admiral, who died here 14th December, 1624, as did his brother, Sir William Howard, in 1600, and whose second son, Charles, afterwards third Earl of Nottingham, was born here in 1610.

* Harl. MSS. No. 29, p. 161.

† Cobbett's State Trials, Vol. I. p. 1154.

‡ Pat. 34 Eliz., pt. 9, & Pat. 9 Jac., pt. 23, No. 5.

I have not been able to discover the year of John Gage's attainder, concerning whom, the anonymous epitomizer of his son has the following:—"His father, through his great misfortune, was brought into the hard condition of confiscating his whole estate, and long imprisonment, having only his life left him, and that onely upon curtesie, by the king's reprieve after condemnation." Still his son is styled, in the inscription on his tomb at Christchurch, Oxford—"***Filius ac Hæres Johannis Gage de Haling, in Agro Suriensi, Armigeri.***"

The same writer informs us that Sir Henry voluntarily demised to his father "the reversion of a faire estate at Croyden, called Haling-house, waiving all respect of wrong to himselfe, or pregudice to his wife and children*."

John Gage, Esq., F.S.A., of Lincoln's Inn, a descendant of the above John Gage, informs us that "he suffered great hardships for the catholic faith, and was long in confinement for harbouring George Beesley, a missionary priest†."

The manor appears again to have become the property of the Gage family, one of whom alienat-

* *Alter Britanniæ Heros*, or the life of the Most Honorable Knight, Sir Henry Gage, late Governor of Oxford, epitomized. Oxford, 1645, page 15.

† Gage's Hist. and Antiq. of Hengrave, 1822, p. 231.

ed it to Christopher Gardner, Esq., in 2 Charles I., in whose family it remained till 1707; when they conveyed it to Edward Stringer, Esq.; whose widow marrying ——— Parker, Esq., her grandson, William Parker Hammond, Esq., inherited it; whose son, W. Parker Hammond, Esq., is the present possessor.

The fine grove in this park contains a great number of exotics and evergreens, which have been celebrated by the laureate Whitehead, in his "Epistle from a Grove in Derbyshire to a Grove in Surry," and "Answer."

Of the "Bourn," which runs by this estate, Camden has written the following:—"For the torrent that the vulgar affirm to rise here sometimes, and to presage dearth and pestilence, it seems hardly worth so much as the mentioning, tho' perhaps it may have something of truth in it*."

No Courts are held for this or the last-mentioned manor.

* Gibson's Camden's "Britannia," 1695 p. 159.

The three following manors are now included in that of Croydon.

PALMERS (*or* TYLEHURST).

By an inquisition in 1595, it appears that Richard Forth, LL.D., died seised of this estate*, which is situated on the south skirts of the Norwood hills, and comprises about seventy acres; it was afterwards the property of the Newlands; the co-heiresses of which family sold it to Mr. Bulkley in 1769†, who disposed of it to Mr. Cotes.

At the time of the inclosure, Mrs. Cotes claimed and had an allotment for the estate as a *farm*.

HAM.

This estate, a farm, situate at the extremity of this parish towards Beckenham, was, in the 2 Philip & Mary, granted by the crown to Antony, Viscount Montague, by the name of "The manor of Estham, *alias* Escheam, *juxta* Croydon," being then part of the honor of Hampton Court.

* Harl. MSS. No. 756, p. 237.

† Horne, p. 310.

Selhurst.

This estate, situate about two miles from the town, on the road to Sydenham, was granted by Henry VIII., in 1541, to Archbishop Cranmer*.

Croydon Park (*now* Park Hill)

Was held by the see of Canterbury, till the reign of Henry VIII., when Archbishop Cranmer exchanged it with that monarch for other lands†; but it reverted to the archbishop by another grant in the reign of Edward VI.‡. In 1326, the keeping of this park was given by Archbishop Reynolds to one —— Le Barber§ for life. In 1382, Sir William Walworth was appointed keeper by Archbishop Courtney‖. In 1405, Richard Hembridge received the same office from Archbishop Arundell¶; and in 1441, Archbishop Chichele granted it to Adam and Richard Pykman**. In the reign of Edward IV., we find its keeping in

* Terrier of Lands in Surrey, Brit. Mus. 4705; Ayscough's Cat.

† Grants of Land and Exchanges, May 4; 31 Hen. VIII., Aug. Office.

‡ Grants of Land and Exchanges, June 12; 1 Ed. VI., Aug. Office.

§ Reg. Reynolds, fol. 261 b.

‖ Reg. Courtney, fol. 37 a.

¶ Reg. Arundel, fol. 401 a.

** Reg. Chichele, fol. 239 a.

the hands of John Lyttyll*, and in the reign of Charles I., Francis Lee, gentleman, held the same by patent, granted by Archbishop Laud, 25th November, 1637†. A Francis Lee, son of Francis Lee, gentleman, of Streatham, had a like grant from Archbishop Juxon, 20th May, 1663‡.

In the time of Archbishop Grindall, Sir Francis Carew, Knt., and one George Withers, had several interests in this park, for redemption of which, the said archbishop paid them the sum of 83*l.* 6*s.* 8*d.*§.

On the sale of the palace, it was in contemplation to erect here a new residence for the archbishop, but Addington being preferred, an act of Parliament was obtained in 1807, for purchasing the mansion and estate of Alderman Tricothick; on the site of which arose the present archiepiscopal seat.

* Excerpta ex computis Ministrorum. Vide Appendix.

† Cart. Miscel., Vol. XIII. No. 16.

‡ Harl. MSS. No. 3797, p. 27.

§ Strype's Life of Grindall, p. 286.

CHAPTER IV.

Addiscombe.

ADDISCOMBE-HOUSE, formerly called Adgecome and Adscomb, but more anciently by its present name, is situate about a mile and a half from the town, on the road to Wickham, and was formerly the residence of a family of the name of Heron. Thomas Heron, Esq., succeeded, in 1514, to the property of his father, an opulent citizen of London, and most probably erected the *ancient* mansion of Adgecome. Here he died in Oct. 1544, leaving issue by his wife Elizabeth, daughter and co-heir of William Bond, Esq., clerk of the green cloth, three sons:—William, justice of the peace for the county, who inherited the estate and died without issue, Jan. 4th, 1562; Sir Nicholas, heir to his brother, who died Sept. 1st, 1568; and Thomas. To Sir Nicholas, succeeded his eldest son, Capt. Poynings Heron, who had, *inter alia*, a daughter baptized at Croydon in 1579. It was afterwards the residence of Sir John Tunstall, Knt., gentleman usher and esquire to Queen Anne, consort of James I., and justice of the peace for the county, whose eldest son Henry, then residing at Croydon, was one of the commissioners appointed in 1647, for inquiring into the conduct of the clergy of Surrey. Sir John died in February, 1650, and his son in the August following, leaving John his son and heir, who, if he resided here, removed before 1662, as the estate was then the property and residence of Sir Purbeck Temple, Knt., of the privy chamber in ordinary of Charles II., who died 29th Aug. 1695, and whose widow dying in Feb. 1700, left it to her nephew, William Draper, Esq., who married Susanna, daughter of the celebrated John Evelyn. This gentleman rebuilt the mansion, commencing in June, 1702, and finishing towards the close of the following year, as we learn from the following extracts from the amusing Diary of his father-in-law.

"27 June, 1702—I went to Wotton with my family for the rest of the summer, and my son-in-law, Draper, with his family came to stay with us, his house at Adscomb being new building*."

"11 July, 1703—I went to Adscomb, 16 miles from Wotton, to see my son-in-law's new house, the outside to the covering being such excellent brick-work, cased with Portland stone, with the pilasters, windows, and within, that I pronounc'd it, in all points of good and solid architecture, to be one of the very best gentlemen's houses in

* Evelyn's Memoirs, ed. Bray, Vol. II. p. 77.

Surrey, when finished. I returned to Wotton tho' weary."* (He was then 83).

The estate became afterwards the property and residence of Charles Clarke, Esq., of Ockley, Surrey, through an heiress of the Draper family; whose only son Charles died in his father's life-time, leaving issue, Charles John, unfortunately killed at Paris by the fall of a scaffold at a public show, on the celebration of the peace of Amiens; and Anne Millicent Clarke, who marrying Emilius Henry Delmè, Esq., afterwards Radcliffe, master of the stud to George IV. and his present Majesty, the estate became the property of

* Evelyn's Memoirs, Vol. II. p. 80.

The following three extracts, in which Addiscombe is referred to, are also taken from Evelyn's Memoirs—

"19 July, 1695—I dined at Sir Purbeck Temple's, neare Croydon; his lady is aunt to my son-in-law Draper; the house is exactly furnished." (Vol. II. p. 48).

"29 August, 1695.—Very cold weather.—Sir Purbeck Temple, uncle to my son Draper, died suddenly. A greate funeral at Adscombe. His lady being owne aunt to my son Draper, he hopes for a good fortune, there being no heir." (Ibid. p. 49.)

"13 February, 1700—I was at the funerall of my Lady Temple, who was buried at Islington, brought from Adscomb, neare Croydon: she left my son-in-law, Draper (her nephew), the mansion-house of Adscomb, very nobly and completely furnish'd, with the estate about it, with plate and jewels to the value in all of about 20,000*l.*; she was a very prudent lady, gave many greate legacies, with 500*l.* to the poore of Islington, where her husband, Sir Purbeck Temple, was buried, both dying without issue." (Ibid. p. 68.)

that gentleman. Lord Grantham, and the first Lord Liverpool, made Addiscombe their place of residence, the latter of whom had a lease of the estate for life; and Lord Chancellor Talbot is said to have resided here.

In 1809, this estate was purchased of Mr. Delmè Radcliffe by the Hon. East India Company, for the purpose of establishing here their military college, previously formed at Woolwich Common, for the education of cadets for the engineers and artillery; which, in 1825, was extended to the reception of the whole infantry service. The cadets, whose number extends from 120 to 150, are under the inspection of an officer of rank in the Company's service, assisted by an officer of His Majesty's corps of engineers or artillery, who examine them previously to their obtaining commissions. There are fourteen professors and masters: teachers of fortification, artillery, engineering, and military tactics in general, mathematics, military and other drawing, lithography, surveying, the classics, the French and oriental languages, chemistry, and geology.

Two public examinations are held annually, at which the chairman and deputy chairman of the East India Company preside, assisted by some of the principal officers of state, there are two terms in the year, one commencing the 1st February and extending to the 16th June, and the other from 1st August to 21st December. The age of the candidates for admission must not be under 14 or exceeding 18 years. The terms are 65*l.* the first year, and 50*l.* for each of the two succeeding years.

Addiscombe house is supposed to have been built after a design of Sir John Vanburgh, and the walls and ceilings of the staircase and saloon to have been painted by Sir William Thornhill. It has since been greatly enlarged by the addition of several unconnected buildings.

On the front is the following inscription in Roman capitals.

NON FACIAM VITIO CULPAVE MINOREM.

CHAPTER V.

Charitable Institutions.

Ellis Davy's Alms House.

This religious foundation is a small unassuming structure situate near the church, and was rebuilt about sixty years since. It was founded on the 27th April, 1447, by one Ellis Davy, citizen and mercer of London, for the support of seven poor people, male and female, including a tutor. He endowed the same with 18*l.* *per annum*, with the addition of four cottages situate near it, the rents of which were to be applied to its repairs.

Having obtained letters patent from Henry VI., dated 25th December, 1445, together with letters patent from Archbishop Stafford, dated 17th February, 1442, and *letters* from the Abbot and convent of St. Saviour, Bermondsey, dated from their Chapter-house, 20th December, 1445, the said Ellis Davy founded this alms-house the 27th of April, 1447—He appointed the vicar, churchwardens, and four of the most worthy householders and parishioners of the town, and their successors, governors of the Alms-house, and constituted the master and wardens of the Mercers' Company, for the time being, overseers of the same.

The said Ellis Davy ordained the tutor and poor of his Alms-house to attend service daily in the church of Croydon, and there to pray upon their knees, for the King, in three Paternosters, three Aves, and a Credo, "with special and hartily recommendation" of the said founder to God and the Virgin Mary. They were also required to say for "the estate of all the sowles abovesaid" daily at their convenience, one Ave, fifteen Paternosters, and three Credos; and after his death, provided he should be buried at Croydon, they and their successors were required to appear daily before his tomb, and there to say the psalm "De Profundis," or three Paternosters, three Aves, and a Credo. He required that their clothes should be "darke and browne of colour, and not staring, neither blazing, and of easy price cloth, according to their degree."

These statutes (which are inserted in full in the Appendix) becoming antiquated, were revised by Archbishop Parker, August 6th, 1566.

The revenue of this charity is, at the present time, 179*l.* 4*s.* 2*d.* *per annum.*

WHITGIFT'S HOSPITAL.

This "memorable and charitable structure," incorporated in the name of the warden and poor of the hospital of the Holy Trinity, was founded in the reign of Elizabeth, by Archbishop Whitgift, for the maintenance of a warden, schoolmaster, and twenty-eight men and women, or as many more under forty as the revenues would admit. The archbishop, having obtained letters patent, with licence of mortmain, from the queen, dated 22 November, 1596, commenced this building on the 14th February, the same year, and finished it 29th September, 1599, expending on the whole, 2716*l.* 11*s.* 1*d.*, as appears from the accounts of the Rev. S. Fynche, vicar of Croydon, appointed by the founder to superintend the works*.

He appointed to the warden a salary of 11*l. per ann.;* and to the schoolmaster, who is also chaplain, a salary of 20*l. per ann.;* and to each poor brother and sister, whose respective ages must not be under sixty, the sum of 5*l. per ann.;* besides wood, corn, and other provisions.

The members he required to be selected—***first,*** from the household of the archbishop—***secondly,*** from the parishes of Croydon and Lambeth—and *lastly,* from such parishes in Kent whose benefices are annexed to the see. The number of the women not to exceed the half of the men, exclusive of the warden and schoolmaster.

The schoolmaster, he required to read public prayers, morning and evening, in the chapel of the Hospital, on all working days except Wednesday and Friday in the forenoon, and Saturday in the afternoon—and to be proficient in Greek and Latin, as also a good versifier of these languages.

He ordained the Archbishop of Canterbury for the time being governor and visitor of the hospital. This trust was delegated by Archbishop Laud, 11th August, 1634, to Sir Edmund Scott,

* Vide "The particular account of the Building of Trinity Hospytall in Croydon, "Lamb. MS. Lib. No. 275.

Knt., and Rev. S. Bernard, vicar of Croydon. For the articles ministered, vide Appendix, No. 15.

He reserved to himself during life the two chambers over the inner gatehouse, and the chamber over the hall, now occupied by the warden; and here he often entertained his noble friends the Earls of Shrewsbury, Worcester, and Cumberland, Lord Zouche, the bishop of London, "and others of near place about her Majesty."

Whilst digging the foundations, several skeletons were discovered by the workmen, of which Mr. Fynche gives an account in two letters to the archbishop. These were in all probability the remains of the unfortunate Londoners who fell in this town on the 14th May, 1624*.

On the 10th July, 1599, between the hours of eight and twelve, the chapel of the Hospital was consecrated by Richard (Bancroft), bishop of London, by the name of "The chapel of the Holy Trinity;" in the presence of Antony, bishop of Chichester, Thomas Montford, D.D., preacher on the occasion, and many others.

Scarcely, however, had he completed this erection, when his enemies, desirous of obtaining a curtailment of the archiepiscopal income, that they

* Vide ante, p. 17. In 1814, Mr. Turner, veterinary surgeon, when digging a gravel-pit in his paddock opposite the Hospital, discovered a number of skeletons, lying about four feet deep from the surface.

might feast upon the spoliation, talked much at Court of the great wealth accumulated by the prelate under his preferment, and of the overgrown revenue of the see. Upon the receipt of this slander, the archbishop immediately drew up a paper, giving the true yearly value of the archbishopric, with an account of all his purchases since his translation. From which paper given by Strype in his life of this prelate, we have extracted the following items.

"These following are for my hospital:—

"The Checker in Croydon cost 200*l.*

"A tenement joining it, cost 30*l.*

"Another tenement in Croydon, called Stay-Cross, with one acre and a half, cost 80*l.*

"Upon these I have builded my hospital, school-house, and school-master's house, and therefore are not rented.

"One piece of ground, called Clotmead, in Croydon, cost 14*l.*, rent 20*s.*

"The Swan in Croydon, *cum pertinentiis*, 80*l.*, rent of this with certain parcels belonging to the Checker, is 13*l.* 6*s.* 8*d.*

One piece of wood land, and some pasture, containing in the whole 77 acres, in Croydon, cost 375*l.*, rent 20*l.*

"One other piece of wood land and pasture in Croydon, cost 410*l.*, rent 23*l.*

"Three other several farms in Croydon cost 1400*l.*, rent 48*l.*"

Of the wonderful condescension of this excellent prelate to the inmates of his hospital, we are told by Izaak Walton, in his life of Hooker, that he visited them so often "that he knew their names and dispositions, and was so truly humble, that he called them brothers and sisters; and whenever the queen descended to that lowliness to dine with him at his palace at Lambeth, which was very often, he would usually the next day shew the like lowliness to his poor brothers and sisters of Croydon, and dine with them at his hospital, at which time you may believe there was joy at the table *."

The same author has also recorded a saying of Boyse Sisi, ambassador from France at the time of the archbishop's death—"The bishop" said the Frenchman, "had published many learned books; but a free-school to train up youth, and an hospital to lodge and maintain aged and poor people, were the best evidence of christian learning that a bishop could leave to posterity †."

Sir George Paul also mentions the many visits of the pious founder to the poor of his hospital.

The revenue, originally only 185*l.* 4*s.* 2*d.* *per annum*, has been greatly increased by fines upon the renewal of leases, chiefly through the care

* Walton's lives, Major's ed. p. 208.

† Ibid.

and attention of the archbishops, Secker and Moore; which, with sundry benefactions, amounted together, in 1817, to 481*l.* 9*s.* 4¾*d.*; and has increased at the present time to 2,007*l.* 19*s.* 4*d.* *per annum.* There are now thirty-four brothers and sisters supported by this charity.

The hospital is a handsome brick edifice, of the Elizabethan style of architecture, in the form of a quadrangle, and situate at the entrance of the town, having, over the entrance, the arms of the see of Canterbury, surmounting the following inscription:—

QVI DAT PAVPERI NON INDIGEBIT.

In the chapel, a small unassuming structure, forming the south-east angle, are preserved the following, among other items:—

* For the following note, I am indebted to the Rev. J. C. Bisset, Chaplain of the Hospital.

By far the chief part of the revenue of the Hospital is derived from the original endowment made by its pious Founder, Archbishop Whitgift, in land and tenements in Croydon, and its neighbourhood. Some estates were afterwards bestowed by other benefactors, who completed the annual income the archbishop designed for the poor in his hospital, and thus contributed to advance its permanent interests. For a long series of years, it was usual for the Hospital to receive a reserved rent for the lands and tenements, on leases renewable every seventh year, upon payment of a fine: that custom has been departed from, and a fixed rent substituted in lieu of all fines. The result of which measure is, that the present rental of the Hospital may be stated as amounting to the sum of 2,007*l.* 19*s.* 4*d.* a-year.

A fine portrait of the founder painted on board, and inscribed above:—

Feci quod potui; potui quod, Christe, dedisti:
Improba, fac melius, si potes, Invidia.

beneath:—

Has Triadi Sanctæ primo qui struxerat ædes,
Illius in veram Præsulis effigiem.

A portrait of a lady in a ruff, inscribed, A.D. 1616.

A frame containing the following:—

TO THE HAPPIE MEMORIE OF Y^E MOST REVEREND FATHER IN GOD,
DOCTOR JOHN WHITGUIFT,
LATE ARCHBISHOP OF CANTERBURIE, ETC.,
HIS GRACE'S SOMETIME FAITHFULL LOVING SERVANT
& UNWORTHIE GENT. USHER, J.W., CONSECRATS
Y^E TESTIMONIE OF HIS ANCIENT
DUTY.

OBIIT 29 *FEB.*, 1603.

Pure Saints by heav'n refyn'd from earthlie drosse,
You duelie can esteeme your new increase:
But our soules' eyes are dymme, to see the losse,
Great Prelate, wee sustayne by thy decease.

Wee never could esteeme thee as we ought,
Although the best men did thee best esteeme;
For hardlie can you fynde a mortall thought,
That of so great worth worthilie can deeme.

This straight sound Cedar, new cut from y[e] Stemme,
  As yet is scarselie myst in Libanus;
This, richer than the wise King's richest gemme,
  New lost, as yet is scarselie myst of us.

But tymes to come, and our deserved want,
  I feare, will teache us more and more to prize
This matchlesse Pearle, this fairest knotlesse Plant,
  On whose top Vertue sitting touch't the Skyes,

Presuming Horace, Ovid confident,
  Proudlie foretold their Bookes' Eternities:
But if my Muse were like my Argument,
  Theis lynes would outlive both their memories.

For their best Maister-Peeces doe contayne
  But Pictures of false Gods, and man's true faults;
Whereas, in my Verse ever should remayne
  A true Saint's praise whose worth fills Heaven's great Vaults.

Shyne bright in y[e] triumphant Churche, faire Soule,
  That in the Militant has shyn'd so longe:
Let rarest witts thy great deserts enrolle,
  I can but sing thee in a mournefull Songe.
And wish, that with a Sea of teares, my Verse
Could make an Island of thy honor'd Herse.

L'Envoy.

Candish in prose sett *Cardinal Wolsey* forth,
  Who serv'd him in that place I serv'd this lord:
He had his faults to write of and his worth,
  Nothing in this man was to be abhorr'd.
Therefore his theme was larger much than mine;
But, *Candish*, my theme better is than thine.

Persius. Helicondasq., pallidamq. Pirenen
Illis remitto, quorum imagines lambunt
Hederæ sequaces. Ipse semipaganus
Ad sacra Vatum carmen affero nostrum.

[Prolog. 5.]

Another frame containing the following :—

ELEGIA

Continens brevissimam descriptionem miserarium et calamitatum
Generis humani quodq. sit subjectu. morti, et quæ tandem ei consolatio.

VITA quid est hominis, nisi . . . . . plena malorum
  Principio, medio, fine dolenda suo.
Cura, labor, morbus, cui, mentem, membra, dolorem,
  Multa, frequens, ... , alit.
Et nihil est aliud, Cæsar, nisi pulvis et umbra,
  Umbra brevis, velox amnis, et aura levis.
Præterit aura levis, velox cito labitur amnis,
  Umbra brevis fugitat, pulvis inanis abit.
Sic hominis properans ævum fugientibus horis
  Labitur, et pluma ceu volitante volat:
Sic homines miseri, quasvis mutamur in horas,
  Et centum vicibus subdita turba sumus.
Ante potes frondes silvarum, germina ruris,
  Et nitidi stellas dinumerare Poli:
Omnia quam numeris valeas includere justis
  Ipse quidem, fragilis massa, laboret homo.
Omnibus incumbit quoddam grave pondus, et omnis
  Vita malis plena est, plena dolore simul.
Et quicquid spirat, vel in aere, in æquore, terra,
  Nil adeo fragile est sicut inermis homo.
Dic mihi de cunctis hominem mortalibus unum,
  Cui non sit sortis certa querela suæ:
Sit licet ille status felicis, et ordinis ampli,
  Sive sit exiguæ conditionis homo;
Sive puer timidus, pueros incommoda mille,
  Mille premunt noxæ, crimina mille premunt;
Seu fueris juvenis, juvenum quoque tempora dura,
  Hic gravis est sudor, perpetuusque labor;
Sive senex tremulus, non ipsa beata senectus,
  Fœta sed est morbis, tristitiaque gravis.

Fac jam sis dives, fortuna est lubrica certe,
Te nunc, nunc alium, nunc aliumq. petit;
Fac sis pauper homo, nescis, O dives amice,
Pauperies secum quod grave portet onus:
Junge tibi uxorem, quæ non miseranda vorabis,
Fert tibi libertas vendita triste jugum;
At maneas cælebs, miser est homo solus et orbus,
Auxilio dulci, subsidioq. caret;
Suscipe discipuli partes parviq. scholaris,
Suscipies natibus multa ferenda tuis;
Vel doceas doctor mitis pia dogmata Christi,
O gravis hæc quantum functio mentis erit;
Sume Senatoris porro tibi munus agendum,
Tum pariter curæ sunt tibi mille datæ;
Fungeris officio regis, non tædia desunt,
Discruciatq. animum plurima cura tuum.
Quicquid agas tandem, nihil est, nihil undiq. tutum
Sis quodcunq. velis, sunt tua damna tibi.
Ut referam verbo, vita est sentina malorum,
Cumq. dolore labor, cumq. labore dolor.
Insuper inconstans hominum est et lubrica vita
Ut cito, quæ speres posse manere, ruant.
Qui modo sanus erat, nunc lecto ægrotus adhæret,
Pauper et est subito, qui modo dives erat.
Cumq. homo proponit, disponunt invida fata,
Cumq. homo vult illic ire, redire jubent.
Cumq. videtur homo pulchram sibi ducere vitam,
Mors venit et celeres injicit ipsa manus.
Quiq. hodie vivit curæ securus inertis,
Verbaq. cum sociis ludet amica suis,
Cras moritur, pharetra tristem portatur ad urnam,
Ut sua det gelido membra tegenda solo:
Tales nosq. sumus, talis nos exitus omnes,
Quotquot in hoc vasto vivimus orbe, manet.
Non multos vixisse dies et sæcla juvabit,
Certa venit tandem funeris hora tibi.

Ipse licet videas longævi Nestoris annos,
Sive tuos Deli, sive Sybilla tuos.
Nil juvat immotus cunctis stat terminus ævi,
Et tandem mortis nigra terenda via est,
Communemq. viam nos ibimus, ibitis, ibunt.
Hic, is, ille, puer, fœmina, virgo, senex,
Mors etenim pede certa ferox venit omnibus æqua,
Quiq. relinquetur non erit unus homo.
Sis ubicunq. velis non evitabile fatum,
Te sequitur terra, te sequiturq. mari.
Hic nil juris habet pauper, cum divite juris
Sive bonus fueris, nil tibi, sive malus.
Occidet infelix, magno cum Cæsare, pastor,
Et nihil hic quemquam caula vel aula juvat.
Occidet afflictus, saturo cum divite, pauper,
Hic nil pauperies divitiæq. juvant.
Omnia sic minuit fatum, sic omnia tollit,
Et simili cunctos sub juga lege trahit.
Atq. humiles altis, imbelles fortibus æquat,
Obscuris celebres, supplicibusq. feros.
Non genus, aut dotes animi, nec respicit annos,
Nec precibus flecti, nec pietate potest.
Illa tamen nobis spes indubitata relicta est,
Quam decet immota nos retinere fide,
Quod licet illa caro, cutis, et qua membra teguntur,
In cava, defuncto corpore, busta cadant.
Non tamen, in tumulis æterna nocte latebunt
Ut caro brutorum, non reditura, jacent.
Sed Deus, ex tumulis, homines educet in auras,
Et rursus veteri vestiet ossa cute.
Id spondent nobis sanctorum carmina vatum,
Et verbum verum, maxime Christe, tuum;
Id rata verba sonant, his nos quoq. credere fas est,
Sydera si cupimus scandere celsa Poli.

SOLI DEO GLORIA.

On the outside, above the window of the chapel, in which is the founder's arms, is the following inscription on a Portland stone:—

EBORACENCIS*
HANC FENESTRAM
FIERI FECIT,
1597.

In the hall, which is situate at the north side of the inner porch, and where the poor brethren dine, is a folio bible, in black letter, with wooden covers mounted with brass, having this inscription:—

Pauperibus Hospitalis in villa de Croydon
Sacrosanctam Trinitatem colentibus
Hoc Verbum Vitæ donavit
ABRAHAMUS HARTWELL*
Reverendissimi Fundatoris
Humilimus Servulus,
1599.

* Probably Michael Murgatroid, the founder's secretary, who is designated "Eboracencis" in his epitaph, vide post, p. 173.

† Abraham Hartwell, M.A., was rector of Stanwich, Northamptonshire; he was secretary to Archbishop Whitgift, and author of "Regina Literata, &c. 1565." "A Report of the kingdome of Congo, a region of Africa; and of the countries that border round about the same," translated from the Italian of Philippo Pigafetta, 1597; "A true discourse upon the matter of Martha Brossier, of Romorantin, pretended to be possessed by a devill," translated from the French, 1599; and "The Ottoman Description of the Empire and Power of Mahomet," translated from the Italian of Lazaro Soranzo, 1603.

Under the inscription is this memorandum:—

Repaired at the expense
of
Thomas Lett, esq.
of Lambeth,
in the year MDCCCXIII.

There were also formerly three antique wooden goblets, one of which, holding about three pints, had the following quaint inscription:—

What, Sirrah! hold thy pease,
Thirste satisfied, cease.

Above the outer gate, in an upper room, called the treasury, are preserved the several papers relating to the Hospital, as purchase-deeds, leases, licences, &c.; of which the Queen's original grant to the founder and the archbishop's deed of endowment are singularly beautiful.

Over the inner gate is this inscription:—

RESTORED 1817; FRANCIS WALTERS, WARDEN.

Adjoining the hospital are the school house and the master's residence. Although the founder has expressly said, that "the howse which I have builded for the sayde schoole howse, and also the howse which I have buylded for the schoole-master, shal be for ever imployde to that use onlye, and to no other;" yet the former is now appropriated to the children of the national school. The latter is still devoted to its original purpose, being the residence of the chaplain.

"This memorable and charitable structure of brick and stone," says Strype*, "one of the most notable monuments founded in these times, for a harbour and subsistence for the poor, together with a fair school house for the increase of literature, and a large dwelling for the schoolmaster, the archbishop had the happiness, through God's favourable assistance, to build and perfect in his own life-time. And the reason why he chose to do it himself while he was alive, was, as Mr. Stowe the historian had heard from his own mouth, because he would not be to his executors a cause of their damnation, remembering the good advice that an ancient father (S. Gregory) had left written to all posterity, 'Tutior est via, ut bonum, quod quisquis post mortem sperat agi per alios, agat, dum vivat ipse, per se;' *i. e.* The good that any one hopeth will be done by others after he is dead, that he do it himself while he is alive, is much the safer way."

* Life of Whitgift, 1718, p. 533.

CHAPLAINS OF ARCHBISHOP WHITGIFT'S HOSPITAL, FROM THE FOUNDATION.

1600 Ambrose Brydges.

1601 John Ireland.

1606 Robert Davies, or Daires, who was deprived.

1616 William Nicolson.

1629 John Webbe.

1651 Thomas Gray.

1668 William Crowe, of Caius Coll., where he matriculated December 14th, 1632; he was born in Suffolk, and was author of a catalogue of the English writers of the Old and New Testament, 1659, which has been frequently printed. He hanged himself about the end of 1674*.

1675 John Shepherd†.

* Wood's Athen. Oxon. Vol. II. p. 344.

† Under this gentleman, John Oldham the poet was three years an usher. John Oldham was born August 9th, 1653, at Shipton near Tedbury, Gloucestershire, and admitted of Edmund Hall, Oxford, in 1670, when he graduated B.A. 1674, and about 1675, became usher to the free-school at Croydon. Here he wrote his satire on the Jesuits, which getting abroad, he was honoured with a visit by the Earls of Rochester and Dorset, Sir Charles Sedly, and other persons of wit and distinction. In 1678, Oldham quitted Croydon, and entered the family of Sir William Thurland, as tutor to his two grandsons; and, in 1681, became tutor to the son of Sir William Hickes. He next applied himself to physic, which he soon relinquished for poetry; and repairing to London, became the associate of his contem-

1681 John Cæsar, M.A., afterwards vicar of Croydon.

1711 Henry Mills, M.A., author of "An Essay on Generosity and Greatness of Spirit," was of Trinity Coll. Oxford, where he graduated M.A., 25th June, 1698. He was rector of Dinder, and prebendary of Wells, and served the cure of Pilton with the chapelry of North Wooton, master of the school of Wells, and vicar of Mestham. Mr. Mills was one of the opponents of Bishop Hoadly, in the Bangorian controversy; for which cause he published a pamphlet, intituled "A full Answer to Mr. Pillonniere's * reply to Dr. Snape, and to the Bishop of Bangor's Preface, so far as it relates to Mr. Mills; in which the Evidences given to Dr. Snape are justified, the Bishop of Bangor's Objections answered, Mr. Pillonniere's pretended Facts disproved, and base Forgery detected; as likewise the true Reasons of such malicious Proceedings against Mr. Mills. The whole supported

porary wits, and a votary of Bacchus. He died of the small-pox, December 9th, 1683, at the seat of his patron the Earl of Kingston, at Holme-pierepoint. Dryden, with whom he was acquinted, and who terms him the "Marcellus of our tongue" has consecrated a beautiful Elegy to his memory.

* Francis de la Pillonniere, a converted jesuit in holy orders, had been usher to Archbishop Whitgift's school, and was now tutor to the family of the bishop.

by ample Testimonies of Gentlemen, Clergy, and many others. In a letter to the Lord Bishop of Bangor, by H. Mills, A.M." He died April 12th, 1742.

1742 Samuel Stavely.

1751 John Taylor Lamb.

1774 James Hodgson, rector of Keston, Kent, who resigned.

1801 John Rose, D.D.* some time under-master of the Merchant Taylors' School, in the commission of the peace, and rector of St. Martin Outwich, Bishopsgate.

1812 John Collinson Bisset, M.A., vicar of Addington, on the resignation of Dr. Rose.

Archbishop Tenison's School.

This school, situate at North End, was founded in the year 1714, by Archbishop Tenison, for

* In 1812, complaint having been made to Archbishop Sutton, of the great mismanagement of Dr. Rose, his grace was pleased to institute an inquiry; when it appeared that he had made the hospital his debtor to the amount of 202*l.* 9*s.* 10*d.*, when at the same time he had appropriated the revenues to his own use. An action by the warden and poor was the consequence; and, in November, 1813, the sheriff's jury gave a verdict for the plaintiffs, 762*l.* 15*s.* 7*d.* Dr. Rose resigned his situation in the April following. The proceedings have been printed.

the education of ten poor boys and the same number of girls; now, on account of sundry benefactions, increased to fourteen boys and fourteen girls, with maintenance for a master and mistress. For the endowment of this institution, he purchased a farm and lands at Limpsfield, in Surry, of the then yearly value of 42*l.*, and bequeathed to it by will the sum of 400*l.*, to be laid out in land for the enlargement of the said charity.

The revenues of this institution having greatly increased, being now about 130*l.* *per annum*, arising from land and money in the Funds, and the old school house becoming unfit for the purpose, the present substantial brick building was erected in 1792, on a space adjoining the old house, which was then let by the trustees. The master and mistress have now a joint salary of 50*l.* per annum. Over the door, on a board, is the following inscription:—

"CHARITY SCHOOL

founded for 14 poor boys and 14 poor girls, by Thomas Tenison, late Lord Archbishop of Canterbury, March 25th, 1714. This present school-house was built in 1791 and 1792, with a legacy of 500*l.*, bequeathed by Mr. James Jenner, and also 300*l.* by Mr. William Heathfield, of London, and donations by the Rev. John Heathfield of Northam, in the county of Hertford, and other charitable persons."

CHAPTER VI.

The Palace.

Of the early history of this once sumptuous and kingly palace, now prostituted to servile uses, nothing has descended to us, and but little of its after-time. Camden says "Those that live there tell you that a royal palace stood formerly on the west part of the town, near Haling, where the rubbish [of buildings] is now and then digg'd up by the husbandmen; and that the archbishops, after it was bestow'd upon them by the king, transferr'd it to their own palace nigher to the river*." But this is only idle tradition, and as such we leave it.

As no additional light has been thrown on the obscurity which involves this venerable structure, we have thought it as well, after noting the respective prelates who are known to have resided here†, to annex the interesting "Account of the

* Gibson's Camden's 'Britannia,' p. 159. See also Gale on the Itin. of Antinonus, p. 73.

† Vide Registers of the See, Lamb. MSS. Lib. The registers of the Archbishops Mepeham, Stratford, Ufford, and Bradwardine are lost.

Palace of Croydon," written by that learned antiquary, Edward Rowe Mores, and published by Dr. Ducarel in his history of this town.

ARCHBISHOPS RESIDENT AT CROYDON.

1273 Archbishop Kilwardby—who issued a mandate from this place, dated 4th September, 1273*.

1278 Archbishop Peckham.

1294 Archbishop Winchelsey.

1313 Archbishop Reynolds.

1366 Archbishop Langham.

1367 Archbishop Witlesey.

1375 Archbishop Sudbury.

1381 Archbishop Courtnay—who received his pall with great solemnity in the great hall ("*in camera principali maner. sui de Croydon*") on the 14th May, 1382†.

1396 Archbishop Arundell.

1414 Archbishop Chichele.

1443 Cardinal Stafford.

1452 Archbishop Kemp.

1454 Cardinal Bourchier.

* Archiepiscopi Cant. mandatum pro convocatione apud novum Templum, London. ex reg. Giffard Wigorn. fol. 41. See Wilkins's Concilia, Vol. II. p. 26.

† Reg. Courtnay, fol. 9 a.

1486 Cardinal Morton.

1504 Archbishop Warham.

1533 Archbishop Cranmer. Hume, speaking of the disgrace and subsequent decapitation of the Duke of Norfolk, says, "Cranmer, though engaged for many years in an opposite party to Norfolk, and though he had received many and great injuries from him, would have no hand in so unjust a prosecution; and he retired to his seat in Croydon" [January, 1547].

1555 Cardinal Pole.

1559 Archbishop Parker.

1575 Archbishop Grindall—who, on being urged to resign the archbishopric, petitioned that he might retain this palace, with the several lands appertaining to it. "Croydon house, he said, was no wholesome house, and that, both his predecessor and he found by experience; notwithstanding, because of the nearness to London, whither he must often repair, or send to have some help of physic, he knew no house so convenient for him, or that might better be spared of his successor, for the short time of his life*."

The sum of his petition was to retain the palace, the meadow adjoining, called "Stubbs," Croydon park, and eighteen acres of meadow, ly-

* Strype's Life of Grindall, p. 284.

ing at Norbury*. He died in this palace, 6th July, 1583.

1583 Archbishop Whitgift. Sir George Paul, in his life of this prelate, writes "And albeit the archbishop had ever a great affection to lie at his mansion-house, at Croydon, for the sweetness of the place, especially in summer-time, whereby also he might sometimes retire himself from the multiplicity of business, and suitors in the vacation; yet after he had builded his hospital and his school, he was farther in love with the place than before. The chief comfort of repose, or solace, that he took was in often dining at the hospital, among his poor brethren, as he called them†."

1610 Archbishop Abbot—resided much at this palace, where he died, August 5th, 1633. In 1617, "This archbishop being at Croydon the day the Book of Sports was ordered to be read in the churches, he flatly forbid it to be read there; which King James was pleased to wink at, notwithstanding the daily endeavours that were used to irritate the King against him‡." Archbishop Abbot cut down the timber, which, till his time, completely surrounded the palace. Among the Harleian MSS. we find the opinion of the Lord

* Strype's Life of Grindall, p. 286. † Ibid. p. 112.

‡ Complete Hist. of England, Vol. 2, p. 709; see also Strype's Life of Grindall.

Chancellor Bacon on this alteration. "The archbish. of Canterbury (Abbot) had a house, by Croydon, pleasantly sited, but that it was too much wood-bound, so he cutt downe all upon the front to the highway. Not long after, the L. Chancellor Bacon riding by that way, asked his man whose faire house that was; he told him, my L. of Canterburie's. It is not possible, sayes he, for his building is inviron'd with wodde. 'Tis true S^r, sayes he, it was so, but he has lately cut most of it downe. By my troth (answered Bacon), he has done very judiciously, for before methoughts it was a very obscure and darke place, but now he has expounded and cleared it wonderfully well*."

1633 Archbishop Laud—Upon whose execution the palace and lands were sequestrated, and, after having been leased to Charles, Earl of Nottingham, were offered for sale, when a survey was made for that prupose 17th March, 1646, by Edward Boyer, Esq., and others. This sale, however, did not take place, and the commissioners granted the estate to Sir William Brereton, Bart.† who resided here during the protectorate.

* No. 6395, p. 90.

† This distinguished parliamentary general was the eldest son of William Brereton, Esq., of Honford, in Cheshire, where he was born,

1660 Archbishop Juxon.

1663 Archbishop Sheldon—retired here after the great plague of London, where he died 9th November, 1677.

1715 Archbishop Wake. Dr. Rawlinson, in his additions to Aubry's Topographical Account of Surry, published in 1718, says, "This seat at present is in a very bad condition, insomuch, that the present possessor of the see of Canterbury has demanded 1400*l.* for dilapidations belonging to this house, which is 1280*l.* more than Pole's executors paid, and 'tis probable, that 'tis 1350*l.* more than was paid by Grindall's executors; and this demand is thought the more severe, inasmuch

1605. On reaching his majority he received a patent of baronetcy, and, in 1628, represented his native county in parliament, and again in 1640. On the 18th August, 1642, on the breaking out of the great rebellion, he narrowly escaped falling a victim to the populace, for ordering a drum to be beat in Chester for the parliament. In the same year he received a commission from that power, to arm the county, and to seize the goods and weapons of the *disaffected;* and in June, 1644, was appointed Major-General of the Cheshire forces. On the appointment of the twelve Major-Generals, in 1655, Sir William had the government of Staffordshire, Cheshire, and Lancashire, conferred upon him.

As a reward for his gallant services, he received, besides the sequestration of the archiepiscopal lands of Croydon, the sequestration of Cashioberry, and other lands of Lord Capel, amounting to 2000*l.* per annum, the chief forestership of Macclesfield, and the seneschalship of that hundred. He died April 7th, 1661. For his military achievements, vide Rycraft's England's Champions, Vicar's England's Worthies, Clarendon, &c.

as His Grace is said to have discovered his intention of suing to His Majesty for a royal licence, that, in a legal way, he may be empowered to pull down some of the buildings at Lambhithe, and the buildings at Croydon, these last being situated, as his Grace apprehends, in an ill air *."

1736 Archbishop Potter.

1747 Archbishop Herring, who died here 13th March, 1757.

1757 Archbishop Hutton.

SOME ACCOUNT OF THE PALACE OF CROYDON, BELONGING TO THE ARCHBISHOPS OF CANTERBURY—BY EDWARD ROWE MORES, M.A., F.S.A.†

The capital residence of the archbishops of this see was anciently the palace at Canterbury, situated near their cathedral, and given by King Ethelbert, after his conversion to Christianity, to Augustine and his successors for ever‡.

* Vol. II. page 33.—Vide "The true Copies of some Letters, occasioned by the Demand for Dilapidations in the Archiepiscopal See of Canterbury. Part II. p. 7: London, 1716."

† Author of "Nomina et Insigna gentilitia Nobilium Equitumque sub Edwardo primo rege Militantium;" "History and Antiquities of Tunstall in Kent;" and several pamphlets relating to the Equitable Society. He was also Editor in conjunction with the Rev. William Romaine, of Calasio's Hebrew Concordance; and published a new edition of Dionysius Halicarnassensis' "De Claris Rhetoribus."

‡ Monast. Ang. Vol. 1, p. 18.

But, besides this palace, the archbishops had many other castles, seats, and manors, where they from time to time resided, as their inclinations for retirement or pleasure did direct them.

Of this number was the manor of Croydon, a place which has for many ages belonged to the see of Canterbury, and is now particularly famous for a magnificent palace of the archbishops of that see, which has been greatly improved and adorned by his present Grace [Archbishop Herring].

To deduce this fabric from its original, and to describe in some sort the alterations which have been made in it since its first foundation, is the subject of my present design: an undertaking so circumstanced, that the great distance of the time at which we are to begin, is one of the least difficulties which will attend us in the prosecution of it.

To discover the original of the palace, we must necessarily seek after that of the manor-house of Croydon, out of which the palace hath arisen; and to come at the original of the manor-house, we must begin with that of the manor itself, whose period of existence is well known to have commenced shortly after the Conquest, when the Conqueror, seizing the lands of the Saxons, and distributing them amongst his Norman followers, created a new kind of tenure, and gave birth to that estate which is now called a manor.

Appertaining to the manor was the capital messuage or mansion-house of the lord of the fee (from whose usual abiding therein the very name of manor is derived); which being therefore so essentially connected therewith, must be esteemed coeval with the manor itself, and to have its beginning between the years 1066 and 1087; within which time this manor of Croydon was given by the Conqueror to Lanfranc, Archbishop of Canterbury.

Here we see the original of this sumptuous edifice, which, from a poor and low beginning, has at length, by the munificence of its possessors, arrived to the state of grandeur in which we now behold it.

When we consider Croydon as belonging for many ages to the see of Canterbury, and by that means exempted from those frequent changes of possession so destructive to the evidences of private property, it may seem no very difficult matter to pursue this fabric through its several gradations, and to point out with some degree of certainty the different periods of its increase and splendour; but it has so happened, that an unfortunate loss of those records, which alone could have been serviceable in this respect, has eluded the most strict search that can possibly be made for this purpose; a fatality which seems to be in a manner peculiar to Croydon: since, had it been necessary to describe the palace of Lambeth, of Otteford, or of Maghfield, or perhaps any other seat which formerly belonged to the Archbishop of Canterbury, such a search had been attended with greater success, and more answerable to the labour of the inquirers.

In a former paragraph we have discovered the original of the palace of Croydon in the foundation of the manor-house of Croydon, and have also very nearly determined the time when that manor-house was first erected: the time so fixed upon will direct us to Archbishop Lanfranc as the founder of it; and the purposes for which it was built will necessarily point out to us the same person for the builder. For Archbishop Lanfranc, having received from the King a certain quantity of land here, which was to be parcelled out amongst inferior tenants holding of the archbishop, must of necessity have prepared a place for assembling those tenants at stated times, to perform the rents and services which by their tenure were due to this manor. I shall therefore not scruple to set down Archbishop Lanfranc as the first founder of the palace of Croydon; and, to corroborate an assertion which I think may be supported by good arguments, I shall add the authority of Eadmer, a monk of Canterbury, who lived in the times we are speaking of, and, in his account of Archbishop Lanfranc, tells us that he built much in the vills belonging to the archbishoprick * †.

But of Lanfranc's building nothing is at this time remaining. Thus much we may venture to say of it, that what was by him built was agreeable to the simplicity of that age, and the uses for which it was intended; being rather calculated for the habitation of the reeve, and the occasional reception of those who owed their suit and service there, than for the residence of an archbishop.

In those early ages, the archbishops either led a conventual life in common with their monks, or at most resided in their palace at Canterbury; and at present I know but one archbishop who lived elsewhere, before Baldwin procured for himself and his successors the manor of Lambeth, by an exchange with the Church of Rochester.

After his time, we have some evidence of the archbishops dwelling at their manor-houses, as Archbishop Walter, at Tenham; Archbishop Langton, at Slindon; and Archbishop Boniface, at Mortlake; but then, and for many years afterwards, as it appears to me, the chief use which the archbishops made of these houses was to supply the place of houses of entertainment upon the road, of

* Hist. Nov. p. 9.

† Archbishop Lanfranc rebuilt Canterbury Cathedral.—G.S.S.

which the country was then in a great measure, if not entirely, destitute. To these the archbishop and his attendants repaired, and were accommodated when they journeyed, as we are expressly told*, Archbishop Stratford did on his way from Canterbury to the parliament holden at Westminster, in the year 1341, whither he was summoned to answer to the articles brought against him by the crown.

But archbishop Kilwardby is the first instance I can produce of an archbishop who ever dwelt at Croydon; and his dwelling there may be taken as a proof that this house was then fit for the reception of so great a prelate.

Notwithstanding which, upon his resignation in the year 1278, the houses and castles of the archbishoprick were much out of repair, and cost his successor, Archbishop Peckham, no less than 3000 marks†, some of which were, in all likelihood, laid out at Croydon; for this archbishop resided much there. In his time begins the registry of the see of Canterbury, which affords the earliest mention I have hitherto seen of a chapel in this manor of Croydon, by recording an ordination holden therein, in the year 1283‡.

Archbishop Winchelsea succeeded. With the lat-

* Birchington, p. 38. † Antiq. Brit. p. 297, edit. ult.
‡ Reg. Peckham, fol. 111 a.

ter part of his pontificate, or perhaps with the age of his successor, Archbishop Reynolds, do coincide certain reparations made at Croydon, the particulars of which are mentioned in the minister's accounts of the year*. The roll is imperfect, and the date wanting; but Richard de Fairford was then bailiff, and Thomas de Bunchesham reeve of Croydon. In that year the kitchen and salsary were repaired, the wardrobe boarded, the bake-house and stable, together with the sheepcotes and stalls for oxen, weather-boarded and put into repair. At this time the buildings were all of timber, no other workmen but carpenters being employed about them. In the same roll is another charge on account of the kitchen-garden and the vineyard. Another article I meet with therein, which, though not relative to our present purpose, I cannot pass over unnoticed; thirty cart-loads of coals were bought this year by the bailiff of Burstowe, which cost, the carriage included, 53*s*. 9*d*.

A very slender and broken clue remains to conduct us through the space of almost an hundred years next succeeding; of which time no inconsiderable part is without any memorial whatever. Within this space Archbishop Wytleseye, residing

* Rot. Lacerat. de temp. E. II.

here, celebrated three ordinations in the chapel of his manor of Croydon, in the year 1371 *.

Archbishop Courtney received his pall in the same chapel †. At the time of his accession to the see, the manors of the archbishoprick were in a very ruinous condition. This general decay in the temporalities occasioned a compromise ‡ to be made between the archbishop and the prior and chapter of Canterbury, which was ratified by the king. Thereby it was stipulated, that certain of the places so fallen to decay should be kept up and sustained by the archbishop and his successors in manner as heretofore; and that the others should only be maintained in meet repair for the purposes of economy and husbandry. In consequence of this, a warrant issued from the archbishop to his steward, 8th January, 1382-3, empowering him to take away the houses, stone, and timber from one part of his manors, and directing him to employ the materials in the most advantageous manner for the service of the others §.

But nevertheless Archbishop Courtney was at vast, heavy, and extraordinary charges; and in his last will earnestly recommends his affairs to the king, entreating him to see that his executors be not injured by an exorbitant demand on the

* Reg. Wytlesey, fol. 167 b & seq.

† Reg. Courtney, fol. 6 a.

‡ Ibid. fol. 51 b.

§ Ibid.

score of dilapidations, and that some regard be had to the bad condition in which he found the possessions of the archbishoprick, and the great sums he had expended on that account*.

I have not been so fortunate as to see the compromise made between Archbishop Courtney and the church of Canterbury, or to meet with any particular of the structures which were repaired in pursuance of it. But, supposing that the manor-house of Croydon was of the number, if we consider the circumstance of the pall being delivered here, which might create in Archbishop Courtney a liking to the place, and that this archbishop spent part of every year of his pontificate at his manor of Croydon, we may reasonably enough be led to think that something more was done here thancommon reparations only; and indeed it appears that the mansion-house of Croydon had at this time increased in its buildings and conveniences; for whereas hitherto the archbishops had no more than one chamber whereto they could retire, which was their bedchamber, so that acts are frequently said to be performed *juxta lectum domini* and *ad pedes lecti;* we now meet with a more honourable apartment, a best or spare room, called the chief or principal chamber†,

* Cant. Sacr. App. No. 13 c.

† Reg. Courtney, fol. 9 a.

used perhaps upon more great and special occasions.

But as the mention of this principal chamber occurs rather too early* in the time of Archbishop Courtney for us to suppose that it was built by him, the credit of my conjecture shall not rest upon this particular alone, especially as I have a far less doubtful testimony to produce in support of it—a chapel built by this archbishop† in his manor of Croydon, underneath the privy-chamber near the garden, wherein a special ordination was held 28th May, 1390. This was a small chapel, intended for private use; the other for public uses. The old chapel, as I imagine, which has been so often mentioned, is afterwards, by way of distinction, called "the chapel of the manor of Croydon ‡."

In the pontificate of the same archbishop also was a new granary, with a chamber over it, erected, and a new wall contiguous thereto built towards the church-yard, which wall was repaired in the year 1400 §.

It may give some little insight into the customs of our ancestors, as well as assist us to form a truer notion of the buildings here, if upon this mention of a wall I should describe a method of constructing one used before we had any know-

* Sc. anno 1382. † Reg. Courtney, fol. 251 a. ‡ Ibid.
§ Comp. Adæ Bochers ppos. de Croyd. 1400.

ledge of bricks: a French invention, the use of which was introduced into England by the attendants on King Henry V. during his wars in France. Before this period, our walls (those I mean which were not built with stone) were composed of timbers set into the ground at proper distances, and covered on each side with laths, which again were covered with plaster. This perishable structure was guarded at the top with concave or roof tyles, to defend it as much as might be against external injuries. Such was the wall here mentioned, contiguous to the church-yard; such the garden wall spoken of hereafter, and many other walls about the manor-house of Croydon.

We have now passed through the fourteenth century, and are brought down to the days of Archbishop Arundell, in whose pontificate the first thing I meet with is the computus of Adam Bochers, reeve of Croydon, from Michaelmas 1399, to Michaelmas 1400. I ascribe this roll to the time of Archbishop Arundell; for though Archbishop Walden was possessed of the see of Canterbury in the year 1399, yet I have good reason to believe that Archbishop Arundell was restored to it before the 2nd of October in that year.

In this roll, Bochers, by his attorney, John Pieres*, accounts for a sum of money expended

* Comp. Adæ Bochers.

in building a new stable and a chamber at Croydon, the particulars of which building are very accurately enumerated. This stable is afterwards mentioned under the name of the *new* stable, and sometimes by the name of the *great* stable, to distinguish it from another, which is called the *old* stable. Between these two stables was a wall of lath and plaster, defended at the top with ridge-tiles; and another wall of the same materials connected with the great stable, with an apartment called the privy chamber, both which buildings were built this year.

The same roll informs us of another building which was erected this year also, and, as it should seem, from the foundations. It is called the hall, and is described as situated opposite to the cellar towards the *herbarium;* but the name does not seem to convey any idea of the building, or of the uses to which it was appropriated.

After the new building, I am to take notice of the repairs of this year; which were, the new hanging the great gate of the manor, and the old racks in the old stable; the reparation of the chamber over the granary; a new door to the cellar; and a new door-case of Caen stone; and without doors, the inclosure about the pond of my lord's garden, and the inclosure of the garden itself were amended, and a new hedge was this year made from my lord's park to a spot between the corner of the kitchen and the pond.

(1401) Soon after these alterations, I read that Archbishop Arundell did ordain an oratory within his manor of Croydon*, which has given occasion to surmise that the chapel belonging to this manor was about this time either rebuilt or repaired; but, considering what we have so lately read of Archbishop Courtney, I should apprehend, this oratory was the private chapel built by him, rather than a temporary structure, set apart for the performance of divine offices, at a time when the principal chapel was not in a condition to be used.

The situation of this chapel, built by Archbishop Courtney, we know by the information of an evidence yet remaining†. A word or two may be added concerning the situation of the principal chapel, and other the buildings before mentioned.

I imagine then that the principal chapel stood where the chapel at present standeth. To the south of it was the principal chamber, adjoining to which was the privy chamber, and to the south of these was the chapel built by Archbishop Courtney; I am aware that what I now say may be liable to an objection, as I have before mentioned a wall connecting the great stable and the

* Reg. Arund. I. 327 b. † Reg. Courtn. fol. 312 a.

privy chamber. But I am of opinion, that this was a different apartment, though it is called by the same name, the privy chamber; otherwise we shall be at a loss to account for the difference of expression used in speaking of this apartment; as it is sometimes called the privy chamber at the east end of the stable, and sometimes the privy chamber towards the garden, a distinction which seems to prove the reality of two apartments, each bearing the same name. But should this disposition of the apartments in the manor house of Croydon be contradicted by any thing which shall be discovered hereafter, or shall appear erroneous to a person well acquainted with the place, which I have never seen but in a drawing, I hope a mistake will easily be overlooked in a matter where the most likely conjecturer is the best historian.

And here again, written evidence failing me, the want of it must be supplied by an authority of less weight indeed, yet such as is not totally to be disregarded; I mean the representation of coat armour upon the buildings, a method much in use with our forefathers, as an expedient to preserve to posterity a remembrance of themselves and their benefactions.

Of the arms, which I am informed are now remaining in the palace at Croydon, I shall take notice of those only belonging to the Archbishops

Arundell and Stafford. The former are placed in the guard chamber, which was built by Archbishop Arundell, in the place where before stood the principal chamber, and may, without impropriety, be called the principal chamber, as rebuilt by Archbishop Arundell. The arms of Archbishop Stafford are more than once repeated in the hall, and may be looked upon, in some sort, as a proof that he repaired and beautified it; which I am inclined to think, rather than that he entirely rebuilt it; because there is something in the building which seems to speak the age of King Richard II.: but as from thence, though a sufficient proof that the building is not older, I cannot affirm that it is so old, I forbore to mention it in its proper place.

(1456) I pass on to the time of Archbishop Bourgchier, who, soon after he came to the see of Canterbury, new tyled * this manor house, and the out-houses belonging to it. For other repairs done at this time, a cart load of freestone was fetched from Mestham, but to what use it was applied I find not. This short account comprehends the chief of the repairs made this year at the manor of Croydon, as they are recorded in an imperfect roll of the 34th of King Henry VI.

* Rot. imperf. de ann. 34 Hen. VI.

But I must not forget to mention, that if evidence were elsewhere wanting to prove that Archbishop Stafford resided at Croydon, it might here be found in a carpenter's bill for making my lord's new bed at the manor house of Croydon.

(1466) A record of other large repairs in this archbishop's time may be seen in the Computus* of Ric. Pykman, keeper of the manor of Croydon, from Michaelmas 1466 to Michaelmas 1467. But the nature of these repairs is only to be guessed at from the materials which were used, and the workmen who were employed about them. Of the former, the most remarkable articles are 5500 tiles, and 24lb. of solder; and of workmen who were employed, one tiler and his man 18 days, another 24 days, and a plasterer 4 days.

(1474) A third reparation of the same sort was made at Croydon by the same archbishop, in the year 1474†; then also were new racks and mangers set up in the stables, and the top of the garden wall was covered with tiles.

(1475) The repeated enumeration of very trifling particulars, which a want of better materials only can excuse, is made capable of some little variation, by what the accounts of the next year‡ present

* Ex comp. omnium minist. hujus anni.

† Comp. Joh. Lyttyll, ex rot. gen. 14 Edw. IV.

‡ Rot. general imperf. de anno 15 Edw. IV.

us with. In them we are informed, that besides the common work of tiling the roofs, and soldering the gutters of the manor house at Croydon, some binns for bread were this year made in the pantry, some work done in the pastry, and a cupboard put up in the hall. Over and above these performances, something more worthy our notice does yet remain to be mentioned: the work done this year over the altar in the chapel, for placing the jewels upon. Perhaps it may be too trifling to say, that the most early mention of a dovehouse at this place is that which I meet with in the accounts of this year.

(1485) The next repairs upon record are not particularized: they fall out in the year 1485*, and in the pontificate of Archbishop Bourgchier; but I can say no more of them than that Joh. Lyttyll, then keeper of the manor of Croydon, accounted for lx*s*. iii*d*. laid out in repairs there done.

(1520) I come now to the reign of King Henry VIII. and the pontificate of Archbishop Warham, making an advance of almost forty years, without any mention of the manor house of Croydon. And what I now meet with, is little more than a gross sum laid out upon the repairs of this place, which,

* Comp. omnium min. hujus anni.

considering the time and other circumstances, is not a small one. The exact time when the sum was so expended is uncertain, because the roll* from whence my information is deduced is imperfect, and the date wanting; but I think it was not far from the year 1520. At that time was expended in making four gates, and in wages paid to sundry carpenters, tilers, and other labourers, the sum of x*l*. xv*s*. ii*d*.

This is the last roll which contains any thing to our purpose, and with it I must conclude a very incomplete account of this eminent structure. An account which is sufficient only to shew what might have been done, had more of these records been preserved to us; and that an history of the palace at Croydon was not without reason expected from the account rolls in the archives at Lambeth.

* Comp. oium min. de temp.

Description of the Palace.

Dr. Ducarel, who prefaced his observations upon the building of Croydon Palace with a short dissertation relating to the antiquity of edifices built entirely of brick, was of opinion that such buildings were not to be found in England till the reign of Henry VI. [1422—1482], and that the east and west sides of the great court of this Palace were some of the first of that age. But we are informed by Leland, that, in the time of Richard II., [1377—1399], the town of Kingston upon Hull "was inclosed with diches, and the waul byon, and yn continuance endid and made *al of brike,* as most part of the houses of the town at that tyme was*." And again we learn, that, in the 1 Henry IV. [1399], licence was given to Sir Roger Tentys to embattle and fortify his mansion-house of Hurst-Monceux, Sussex, which is wholly of brick. According to Dean Lyttleton, Sir Roger availed himself of this licence soon after it was obtained. From these facts, therefore, we may be authorized to infer an earlier origin to the east and west sides of the Palace than Ducarel has assigned.

The palace of Croydon, including offices and stables, formed in its perfect state an irregular quadrangle; the interior of which was about 156 feet wide from east to west, and 126 feet from north to south. It is lowly situate, and approached through an avenue from Church Street, once guarded by iron gates—the piers of which, erected by Archbishop Potter, bear the date 1742†. The whole building was of brick, except the guard chamber, the great hall, the kitchen, and adjoining offices, which were of stone.

The demesne extended over a space a little exceeding fourteen acres.

* Leland's "Itinerary," by Hearne, 1710, Vol. I. p. 41.

† At this gate, during the time of Archbishop Herring, the ancient alms, called "The Dole," was distributed.

PORTER'S LODGE.

The gateway and porter's lodge, with housekeeper's* house, forming the north-east angle,

* The Keeper of the palace or mansion-house of Croydon enjoyed his office by patent from the archbishop. The names of those that I have been able to meet with are, Adam and Richard Pikeman, appointed by Archbishop Chichele, in 1441, (Reg. Chichele, Part 1, fol. 239 a); John Lyttyll, in 1474, and again in 1483, (Excerpta ex Computis Ministrorum); Ralph Macon, Richard Toverdine and Mathew Jenkins, the immediate predecessors of Ralph Watts and Sir George Askew, who were appointed by Archbishop Abbot in 1630, (Harl. MSS. No. 3797); Ralph Watts and Paul Wid-

were taken down about 1806, with the exception of the stone arch of the inner gate, and were con-

dop, appointed in 1661, (Harl. MSS. No. 3798); and Edward Starke appointed by Archbishop Sancroft, (Harl. MSS. No. 3797, p. 153). His duty was as follows:—"To be and inhabit in the house above granted him for his own habitason, soe to be neare his busines. wch is very carefully & dilligently to take notice of all defects & want of reparons of the mason house, stables, other outhouses, moates, ponds, pipes, & other watercourses, etc.; and, as any small defects hap^ne^ to be, immediately to take care that they be repayred; but if they hap^ne^ to be any thing considerable, then & in such case, to make his applicason to us, or o^r^ officers, and take & follow such directone as from time to time be given him by us or them for their speedy repay^n^ & amendm^t^.

"He shall take greate care of all the goods & furniture wch now are, or at any time hereafter shall be brought into our said mansonhouse: and of & for wch he shall make himself answerable by subscribing our Stew^ts^ book of household goods; & from it taking an exact coppie signed by o^r^ steward, shall keep it safe, soe to prev^t^ any mistake or misreconing that may arise concerning them.

"He shall especially in most weather take care that surch be made in the severall respective roomes of o^r^ s^d^ mansion-house for ye ayring of it, & the s^d^ household goods, wch will be very necessary, at least in the absence of us and of our successors from that place. And for wch purpose, he shall have a competent allowance of fewell by war^t^ & assignm^t^ out of our woods commonly called Northwood.

"He shall doe his uttermost endeav^r^ to serve to us & our successors all our royalties of hunting, hawking, & fishing, in the man^ers^ of Croydon and Waddon; he having our deputason, giving him power & authority to preserve the game: according to ye act of parliam^t^ in that case made & p'vided.

"He shall take care of all the demeasn lands wch now are, or wch hereafter may be in ours, or o^r^ successors hands. To see that y^e^

sidered by Ducarel to be of the age of Henry VII. In a MS. survey, made in the time of Archbishop Sancroft, it is thus described:—

"It consists off a little dining roome, and a chamber over it, and are within the ffoundacons of the Tower. Over them are other lodging roomes (not belonging to the house-keeper) but were usually the chambers of the chaplains.

"From that upper chamber there is a passage, on the left hand of which is a little chamber; and from that passage, on the right hand, is another passage wch leads to three pretty good chambers with a closset or two belonging to them; and beyond them is a little chamber over the gate, wch leads into the back yard.

"Below stairs, is a convenient kitchen, and a roome by it, constantly made use off for a larder; and on the same ffloor are other little roomes for cellars, and for convenient laying of lumber out of the way.

"Behind the house in a little yard (or curtil-

hedges & fences be kept in good & sufficient rep. & to advantage & benefit of ye s^d^ lands."

There was an ancient fee of twopence per day attached to the office, and the rent of several butchers' stalls set up in the market by Sir W. Brereton. In Archbishop Sancroft's time, I find the salary amounted to 10*l.* *per annum*, exclusive of the stalls, then let for 6*l.*, but inclusive of the fee. (Harl. MSS. No. 3797, p. 153).

lage) is a wash-house; and, beyond that, a small garden paled into the house.

"There are two stables, and over them two hay-lofts; each stable will hold 4 or 5 horses; and stand adjoining the Porter's Lodge*."

The stables, extending from the Porter's Lodge along the north side, have been recently converted into cottages.

The west side of the palace was removed in 1808, for the purpose of enlarging the church-yard†. The doors and windows were narrow at the top; a long passage from a single staircase led to the rooms above, which were square, each having a chimney, and a small window looking into the churchyard.

The east side, extending from the Porter's Lodge to the site of the kitchen offices, differs only from the west in respect to the stairs, this side having several flights, and some of the rooms being provided with a closet, but without any chimney. These apartments were occupied by the constant retainers of the archbishop; while those on the west were appropriated to visitors.

* Harl. MSS. No. 3797, p. 152.

† The site of this building, measuring about one rood, was consecrated the following year by Archbishop Sutton.

INTERIOR OF GREAT HALL.

The great hall, the former banqueting and council chamber of the archbishops, appears to have been built by Archbishop Stafford, whose arms, with those of Humphrey and Henry Earls of Stafford, are to be seen in several places. The porch, which forms the principal entrance to the palace, projects from the north-east corner towards the court, and has the appearance of greater antiquity than the hall, the arches of the doors being in the old mitred style. Over this porch was formerly a small chamber; and opposite is another entrance, leading to the garden.

The length of the hall is 56 feet; the width 37ft. 9in.; and the height, 37ft. 6in. It is lighted on the south by four windows, and on the north by three only, the space allotted for the fourth forming one of the sides to the chamber over the porch. These windows have a depressed arch for their head, and are divided by two vertical mullions, without a transom. On the north is the orielle window, differing from those usually placed in such situations, being here uniform with the rest. In the centre was formerly a fire-place with a louvre above. The music gallery was situate at the east end, and supported by a screen. The louvre was taken away by Archbishop Herring; who also removed the orielle, or long passage, at the west end, and filled up the long, narrow window behind the music gallery. This window extended from the string-course to the roof. Immediately below the window were three arched doors leading to the buttery, kitchen, and cellars. The latter offices connected the hall with the east or servants' apartments, and were taken down about 1810. Ducarel considered this part of the building to be of the age of Richard II.

At the upper end of the hall was, till lately, the following remarkable coat of arms, supported by two angels, viz. *azure*, a cross fleury *or*, between five martlets of the second, for Edward the Confessor; empaling quarterly 1 and 4 *azure*, three fleurs de lis *or*, for France; 2 and 3 *gules*, three lions passant guardant *or*, for England. Under these arms is another angel, holding a scroll, which bears the following inscription, now almost illegible:—

Dne salbum fac regem.

These arms, which may be considered as coeval with the hall itself, originally stood in the orielle or passage before mentioned, from which place they were removed by Archbishop Herring to that just described.

The covered crown, surmounting the shield, being first used by Henry VI., and his arms on the charter of foundation of Eton College having the same supporters, this coat has naturally been assigned to that King.

Directly under this coat, on the string-course, were the arms of Archbishop Stafford—*Or*, in a border engrailed *sable*, a chevron *gules*, charged with a mitre of the first. This archbishop is supposed to have built or entirely repaired the hall. On the north-east corbel were the arms of Humphrey, Earl of Stafford, created Duke of Buckingham 1444, to whom the archbishop was related—*or*, a chevron *gules;* and on the opposite corbel were those of the see of Bath and Wells, incorrectly emblazoned, *azure*, a saltire *or*, being given for *azure*, a saltire quarterly *or* and *argent*.

The whole of the eastern side of the hall fell down on the 8th June, 1830. Upon its re-instatement, the four several coats of arms, which we have just described, were placed at the opposite end.

On the second corbel on the south side are the arms of the see of Bath and Wells, empaling the arms of Archbishop Stafford, which is in this instance without the border. Here again is another error, the dexter side instead of the sinister bearing the arms of Stafford. On the next corbel are the arms of the see of Canterbury, empaling Stafford, as described in the first-mentioned coat. Affixed to the fourth corbel are the arms of the same see, empaling those of Archbishop Herring*, *gules*, semée of cross croslets *argent*, three herrings hauriant of the last. On the north-west corbel, are the arms of the same see, empaling those of Archbishop Laud, *sable*, on a chevron *or*, three cross pattées fitchie *gules*, between three estoiles *argent*.

* Archbishop Herring completely repaired and fitted up this palace, at an expense exceeding 6000*l*. Notwithstanding this large outlay, his executors were sued by his successor, Archbishop Hutton, for dilipidations; and that prelate dying before the suit was terminated, Archbishop Secker recovered damages amounting to 1,564*l*. 4*s*. 11*d*.

On the second corbel on the north side, is an unknown coat; quarterly 1 and 4 *gules*, a chief *or*, 2 and 3 cheque *or* and *azure*, a chief of the second, over all a bend sinister *or*. On the third corbel on this side are the arms of Henry, Earl of Stafford; quarterly, 1, the arms of France and England empaled, in a border *or*, in mistake for *argent*; 2 and 3 *azure*, on a bend cotised *argent*, three mullets *or*, between six lions rampant *gules*, for Bohun; 4 *or*, a chevron *gules*, for Stafford. On the next corbel are the arms of France and England quarterly, with a label of three points, supposed to be the arms of Richard, Duke of York, the leader of the red rose party. On the last corbel, are the arms of the see, empaling Archbishop Juxon's* *argent*, a cross *gules*, between four moors' heads full faced proper.

The Guard Chamber,

Situate on the west of the Great Hall, is 50ft. 8in. long, and 22ft. 6in. wide, having a fireplace in the middle of the north wall, with a frame above, once containing a scriptural or landscape painting. Opposite to the fireplace is a bay window of more modern date than the room itself, looking into a small court.

* This Archbishop also repaired the Palace.

This chamber was probably built by Archbishop Arundell, whose arms, quarterly, 1 and 4 *gules*, a lion rampant *or*, 2 and 3 cheque *or* and *azure*, appear in the north corner, empaled by those of the see of Canterbury; and by themselves in the south corner. The other arms in this room, are those of the Archbishops Laud, Sheldon (*argent*, on a chevron *gules*, three sheldrakes of the first, in a canton of the second, a rose of the last), Juxon, Cranmer (*argent*, a chevron *sable*, between three cranes of the last), and Parker (*gules*, on a chevron *argent*, three mullets of the first, between four keys of the second), the arms of England (quarterly, 1 and 4, France and England empaled; 2, Scotland; 3, Ireland), and those of France and England empaled*.

The long gallery † was rebuilt by Archbishop

* The following arms, which I have not been able to discover, are mentioned by Ducarel as being in Croydon Palace, in 1755.

"1 First and Fourth, *gules* a chief *or;* over all, a bend of the second and third cheque *or* and *azure*, a chief *or*.

2 Cromwell of Lincolnshire, [*arg.* a chief *gu.*, and baton *az.*]

3 Harrington of Derbyshire, [*or* a chief *gu.* on a bend *az.*, an annulet.]

4 Sir Thomas Garen, by the name of Palmer."

The arms of the see, empaling Laud, with the date 1638, are to be found on the wall by the road leading to the old town.

† In the MS. library at Lambeth palace is preserved a pane of

Wake*, on the site of the ancient one, and forms part of the south front of the palace.

The green-house, forming the south-east angle of the palace, has been converted into a dwelling.

The other rooms have been so much altered and otherwise abused, that we pass them over. Ducarel, who considered the dining-room (which extends from the guard-chamber to the church-yard), the adjoining apartments, and the rooms and the offices underneath, as one body of building, says:—

"The dining-room is of brick; the ceilings of some of the rooms underneath are of wood, and very low; the windows below stairs but small; and though they are not of the same make as

glass, taken from one of the windows of this gallery, having the following inscription.—

"Memorand. Ecclię de
Micham, Cheme & Stonę cum aliis
fulgure combustæ sunt
Januarii 14, 1638-9.
Omen avertat Deus."

together with a paper inscribed by Archbishop Wake as under:—

"This Glasse was taken out of the west window of the gallery at Croydon before I new built it; and is, as I take it, the writing of Archbishop Laud's own hand."

* The executors of this prelate were sued by his successor (Archbishop Potter) for dilapidations in the Court of Arches, 1737-8, who recovered 2006*l.* 9*s.* 1¾*d.*, of which 484*l.* 13*s.* was on account of this palace.

those of the east and west sides of the great court, yet I take this building to be near as old, and to have been built some time in the reign of King Henry VI. It hath been so frequently repaired and altered by the several archbishops of this see, that there are at present few or no marks to ascertain the time when it was first erected. I could observe no more than two.

"The first mark ❀ below stairs in stone, in two places over the door at the bottom of the private staircase that goes into the garden.

"The second mark I took notice of, was several smaller roses, thus * *, which I observed on different parts of the ceiling in the apartment adjoining to the great dining-room, which, I take it, was formerly a retiring or withdrawing room. As I do not remember to have seen these roses upon any buildings before the reign of Henry VI., I conclude they were built about that time."

The Chapel

Is situate to the north of the guard chamber. In the registers of the see we find that ordinations were held in the principal chapel (*in capella principali manerii de Croydon*), in the chapel of the manor of Croydon (*in capella manerii de Croydon*); and in Archbishop Courtney's register, is recorded an ordination held in the private chapel beneath the privy chamber, near the garden, "*jam de novo constructa**." Of the private chapel, no vestige now remains; nor have we any mention of it, save in the above instance: the principal chapel, and the chapel of the manor of Croydon, were in all probability the same. The present chapel is certainly not that in which Archbishop Peckham held an ordination, 15th December, 1283†, though most likely erected on the same spot. The papal badge of the Cross Keys, on the western gable, as seen from the churchyard, denote an earlier date than 1559, the year in which the first protestant archbishop was consecrated.

The chapel is an oblong brick building, ascended from the garden and churchyard by a projecting stone staircase at the north-west corner, and had once a small belfrey at the west end. In the interior, the choir is divided from the anti-chapel by a neat screen, having open panels at the top, and but little tracery. There was a window at either end, but that on the west has long been closed. On the north side are three square windows, divided severally into five compartments by four vertical mullions terminating in depressed arches; and on the opposite side, two

* Fol. 251 a. † Reg. Peckham, fol. 111 a.

windows of the same construction looking into the court, situate between this part of the edifice and the guard chamber. The window over the altar is divided by six vertical mullions; it was formerly of stained glass, as appears by the published account of Archbishop Laud's trial, where it is stated, that "Browne, his joiner, being examined at the Lords' bar against his will, confessed upon his oath, that, in the chapel of Croydon, there was an old broken crucifix in the window, which he, by the archbishop's direction, caused to be repaired and made complete; which picture was there remaining very lately; for which work, Master Pryn found the glazier's bill, discharged by the archbishop himself, among other of his papers." The seats, which surround the room, are faced by low screens, ornamented at each end by a rising shield, carved generally on both sides, with the arms either of Laud or Juxon. The arms of Laud are empaled successively by those of the sees of St. David's, Bath and Wells, London, and Canterbury; of the deanery of Gloucester; and of St. John's College, Oxford. The arms of Juxon are also painted on either side of the centres of the arches which support the roof. At the south-west corner is a pulpit, overlooking the screen, and ornamented by the arms of the archiepiscopal see empaling Laud. The archbishop's seat, at the right of the entrance of the choir, is conspicuous, and has a canopy.

Archbishop Laud, it appears, placed an organ in this chapel; for in his will, a copy of which is preserved in the MS. library at Lambeth, is the following legacy:—"Item, to Mr. Cobb, my organ that is at Croydon." He also intended to repair and beautify it throughout, a task which his unhappy death prevented him from completing.

After the execution of this prelate, the palace and estate, as before mentioned*, passed from the hands of the Earl of Nottingham into the possession of Sir William Brereton, Bart., "a notable man at a thanksgiving dinner," to use the words of a writer of those days, "having terrible long teeth, and a prodigious stomach, to turn the archbishop's chapel at Croydon into a kitchen, also to swallow up that palace and lands at a morsel†." In this state the chapel remained till the restoration of the see, when Archbishop Juxon continued the unfinished work of his unfortunate predecessor, as is evident by his arms being placed in several parts of it.

The following bishops were consecrated in this chapel:—

June 26, 1553.—John Taylor, D.D., Bishop of Lincoln, by Arch-

* Page 78.

† The mystery of the good Old Cause briefly unfolded, 1660.

bishop Cranmer, assisted by Nicholas (Ridley), Bishop of London, and John (Scory), Bishop of Rochester*.

May 6, 1553.—John Harley, D.D., Bishop of Hereford, by the same archbishop, assisted by Nicholas, Bishop of London, and Robert (Aldrich), Bishop of Carlisle†.

August 2, 1579.—John Woolston, D.D., Bishop of Exeter, by Archbishop Grindall, assisted by John (Elmer), Bishop of London, and John (Young), Bishop of Rochester‡.

September 18, 1580.—John Watson, D.D., Bishop of Winchester, and William Overton, D.D., Bishop of Lichfield and Coventry, by the same archbishop, assisted as before‡.

September 3, 1581.—John Bullingham, D.D., Bishop of Gloucester, by the same archbishop, assisted as before‡.

August 4, 1628.—Richard Mountague, D.D., Bishop of Chichester, by William (Laud), Bishop of London, Richard (Neale), Bishop of Winton, John (Buckeridge), Bishop of Ely, and Francis (White), Bishop of Carlisle§.

September 7, 1628.—Leonard Mawe, D.D., and Walter Curll, D.D., the first Bishop of Bath and Wells, the latter of Rochester, by Archbishop Abbot, assisted by Richard (Neale), Bishop of Winton, John, Bishop of Ely, and Francis, Bishop of Carlisle‖.

October 24, 1630.—William Peirse, D.D., Bishop of Peterborough, by the same archbishop, assisted by Richard, Bishop of Winton; Theophilus (Field), Bishop of St. David's, Richard (Corbet), Bishop of Oxford, and John (Bowle), Bishop of Rochester**.

* Reg. Cranmer, fol. 335. † Ibid. fol. 583.

‡ Strype's life of Grindall.

§ Laud's Diary, page 3. That the primate did not officiate at this consecration, may be accounted for by the fact, that Archbishop Abbot was about this time suspended for refusing to license the printing of a sermon by Dr. Sibthorp, justifying the King's demand for a loan; or Dr. Mountague might possibly have refused (as was the case with many) to receive consecration at his hands, on account of his unfortunate accident in shooting Lord Zouche's gamekeeper in Leicestershire, 1621.

‖ Reg. Abbot, Part 2, fol. 156. ** Ibid. fol. 23.

As Archbishops Secker and Cornwallis, Hutton's immediate successors, did not make Croydon their place of residence, this palace became greatly dilapidated; and, in 1780, an act of Parliament was obtained "for vesting in trustees the capital messuage, with the appurtenances, at Croydon, in the county of Surrey, known by the name of The Palace of the Archbishop of Canterbury, and two closes near thereto adjoining, in trust, to sell the same; and for disposing of and applying the money to arise thereby, and received on account of the dilapidations thereof, and other money, in the manner and for the purposes therein mentioned."

In the preamble of this act, it is stated—1. That the palace was in so low and unwholesome a situation, and in many respects so incommodious and unfit to be the habitation of an archbishop of Canterbury, that few of the archbishops had of late years been able to reside there, and the same was then unfit to be their habitation.—2. That there then stood on the books of the South Sea Company, 5,402*l.* 3*s.* 3*d.* Old South Sea Annuities, in the names of Frederick, Lord Archbishop of Canterbury, and Richard Maurice Jones, Gent., deceased, in trust for the see, as stock which had been formerly purchased with money allowed by the Commissioners for building Westminster-bridge, as a compensation to the Archbishop of Canterbury and his successors for the loss they sustained by destroying the horse-ferry from Lambeth to Millbank; the dividends whereof, amounting to 162*l.* 1*s.* 2*d.*, had been received by the Archbishops of Canterbury for their own use and benefit.—3. That there was also standing in the name of the archbishop, 1,564*l.* 4*s.* 11*d.* 3*l. per cent.* Consolidated Bank Annuities, purchased by him in May, 1769, with the monies received by him for dilapidations at Croydon, and which, with the accumulated interest, amounted to 2,360*l.* 0*s.* 3*d.*—4. That the archbishop had then lately purchased the leasehold interest in a farm belonging to the see of Canterbury, called Park Hill, situate within half a mile of the town of Croydon, and very proper for building on part thereof a new palace for the use of the said archbishop and his successors, in lieu of the palace at Croydon.

By this act, the palace and appurtenances, described as "an ancient capital messuage, originally intended for the place of residence in summer of the Archbishops of Canterbury," were vested in four trustees, the Lord Chancellor, the Lord Chief Justice of the King's Bench, and the

Bishops of London and Winchester for the time being, with power to sell the same, either together or in parcels, or to pull down the buildings and sell the materials: who, by virtue of such power, sold by auction, October 10th, 1780, to Abraham Pitches, Esq., of Streatham, (afterwards Sir Abraham)—"The freehold and absolute inheritance in fee simple of the said capital messuage or mansion-house, with its rights, members and appurtenances, and also all houses, out-houses, edifices, buildings, gardens, orchards, tenements, hereditaments, and appurtenances, to the said capital messuage or hereditaments belonging, and their and every of their appurtenances; and also two closes of land, containing by estimation about six acres, contiguous to or near the said capital messuage, with their appurtenances; and also the water-conduit or conduits, situate in a mead called Parson's Mead, in Croydon, with the aqueducts, and the leaden pipe or pipes leading therefrom to a cistern in the said palace," for 2,520*l.*

Thus the palace of Croydon, the residence of the primates of all England for so many centuries, —the seat of learning and the scene of unequalled splendour, eventually became appropriated to the printing of linen; whilst its gardens and orchards have been converted into a bleaching ground*.

Qua Troja fuit, nunc est seges!

* Since this work was prepared for the press, the whole demesne has been sold in various lots, the purchasers of which are rapidly converting many portions of this ancient pile into modern dwellings; and its total demolition, which time could not have effected in centuries, will probably be achieved in a few years.

CHAPTER VII.

The Church.

THIS noble monument of the piety of our ancestors, dedicated to St. John the Baptist, is considered one of the finest examples of ecclesiastical architecture in the county; it is situate at the bottom of the town adjoining the palace lands, and is of unknown antiquity. For although its foundation is said to have been laid in the time of Archbishop Courtney, it is merely a conjecture, arising from the arms of that prelate, (*or*, three torteaux), being till lately affixed to the north entrance. And to the herald, we are again indebted, when we place the date of its completion in the days of Archbishop Chicheley*, who expended large sums on its building†, and whose arms, (*argent*, a chevron *gules*, between three cinquefoils of the last), are to be seen, terminating on one side the spandril of the arch over the west or principal entrance.

Domesday book informs us, that there was a church in Croydon at the time of the conquest; and looking still further back, we find that a church stood here in the Saxon era: for, to the will of Byrhtric and Ælfwy, made anno 960—a copy of which is printed in Lambard's Perambulation of Kent—is witness Ælffie, the priest of Croydon.

The present church is a large handsome stone building of the pointed style of architecture. Its lofty square tower, built of stone and flint, is exceedingly well-proportioned, and rises to the height of four stories; it is supported by strong buttresses, and adorned at the summit by battle-

* "Henry Chicheley, Archbishop of Canterburie, was the new builder or especial repairer of Croydon church, as appeareth by his arms graven on the walls, steeple, and porch." Stowe's Annals, p. 631.

† Duck, vita Hen. Chichele, 1617, p. 107.

ments and crocketed pinnacles, issuing from octagonal turrets; it contains a good ring of eight bells, with chimes, which play a psalm tune every six hours.

These bells are inscribed as follows:—

1 My voice I will raise.
And sound to my subscribers' praise
At proper times.—Thomas Lister made me, 1738.
2 Thomas Lister fecit, 1738.
3 Thomas Lister fecit, 1738.
4 T. L. 1738.
5 T. L. 1738.
6 Thomas Lister, Londini, fecit, 1738.
7 Robert Osborn and Francis Meager, Churchwardens. Thomas Lister, Londini, fecit, 1738.
8 Mr. Nath. Collier Vicker, Robert Osborn and Francis Meager, Churchwardens. Thomas Lister fecit, 1738.

On the south-east corner, hangs the Saints' bell, its usual position, bearing this inscription;—

Francis Tirrell gave this bell, 1610. Recast in 1757.

The tower being repaired some years ago, the buttresses were entirely cased with Roman cement, and the ornaments restored, when the following inscription, commemorative of the same, was placed under the fine mullioned window, immediately over the entrance:—

THIS TOWER REPAIRED IN 1807 & 1808, WILLIAM BROWN AND JOHN PHILLIPSON, CHURCHWARDENS.

In regard to the time when a vicarage was founded in this church, we are again at a loss—

its first mention is in 1289, when Henry de la Rye was presented to the vicarage of Croydon by Ægidius de Audenado, rector of the same. About which time, as we learn from a Valor Beneficiorum, compiled 20 Edw. I., and preserved in the Bodleian Library, the rectory was valued at 60 marks, and the vicarage at 15 marks. In 1534, the vicarage was valued at 21*l.* 18*s.* 11½*d.** And in Ecton's Thesaurus Rerum Ecclesiasticarum it is thus stated—

Clear yearly value		Yearly tenths
45*l.* 0*s.* 0*d.*	CroydonV. (St. John Baptist)[Pecul]	2*l.* 3*s.* 10*d.*
	Pens. Prior. de Bermondsey	
	cvi*s.* viii*d.* Redd. Mansion. ii*d.*	

In the Liber Regis, the vicarage, which is discharged from the payment of first fruits, is valued at 21*l.* 18*s.* 9*d.*

An instrument, dated Maidenston, 2nd June, 1348, copied at length in Appendix, contains an ordination held by Archbishop Stratford, to consider what portion of tithes belonged respectively to the rector and vicar of this church. At this ordination, it was settled that the rector should have all the great tithes, viz. corn, hay, falls of wood and timber within the parish, all live mortuaries due at funerals, and a moiety of the tithes of lambs, which are to be tithed *per capita,* with a

* Reg. Winton, Fox, pl. 5.

pension of eight marks to be paid by the vicar in equal portions on the festivals of St. Michael, Christmas-day, Easter, and the Nativity of St. John the Baptist. The vicar to hold the vicarage house and garden, all oblations in the said church, a moiety of the tithes of lambs tithed *per capita*, the money arising by right from lambs not tithed *per capita*, all tithes of wool, calves, pigs, geese, ducks, pigeons, cheese, milk, butter, herbage, apples, pears, and other fruits, with all tithes of flax, mustard, eggs, and merchandize*. The said vicar is also required, with the assistance of another priest, to perform divine service in the said church, and to enjoy the ministering of the bread, wine, and candles; he is also required to find such books, surplices, vestments, and ornaments for the said church as are generally found by the rector or vicar by custom or right, to pay the tenths and all other charges imposed on the

* Ducarel, pp. 109—122, has given three decrees relating to the Vicarial Tithes. The first in the Exchequer, Hil. Term, 31 & 32 Car. 2 (1679-80), between Hatcher and others, plaintiffs, and Dr. Clewer or Cleiver, defendant. The second, at Serjeants' Inn, Hil. Term, 33 & 34 Car. 2 (1681), between Dr. Clewer, plaintiff, and Pullen and Others, defendants. The third, in Chancery, July 2, 1743, between Collier, clerk, plaintiff, and Heathfield and others, defendants. But as the vicarial tithes are now compounded for, I have thought it unnecessary to enlarge the Appendix by the insertion of these decrees.

church of England, according to the acknowledged taxation of 10*l.* sterling, at which the said vicarage is taxed. The said vicar is also required to repair the chancel of the said church, viz. its roof and walls, externally and internally, and to be at the expense of all ordinary and extraordinary reparations required by the said church. The archbishop reserved to himself and his successors the right of augmenting or decreasing the revenues of the vicarage*.

In the 11 Edward II., an "*Inquisito ad quod damnum*†" was held previous to an exchange between Archbishop Reynolds and the prior and convent of St. Saviour's Bermondsey, of the advowson of this church, and two acres of land, of the yearly value of 2*s.*, for a rent-charge in Wichesflete, of 28*l.* 12*s.* 11*d.*, and one hide of land and two mills, with their appurtenances, in Southwark, valued together at 10 marks and 4*s.* *per annum.* The reason of this intended exchange, for it is supposed that it never took place, the instrument being crossed out in the register, and the succeeding archbishops continuing to

* In 1797, when the act of Parliament was obtained for inclosing the parish, a glebe was allotted, under the said act, to the vicar, but the parish is still subject to vicarial tithes, except the further part of Norwood; the right of Easter offerings is expressly continued to the vicar throughout the whole parish without exception.

† Reg. Reynolds, fol. 98 b.

present to the vicarage, was owing to the deterioration of the revenues of the convent by an inundation of water; and the archbishop, fearing lest the said convent should be dissolved, proposed granting them the appropriation of the church of Croydon.

In the register of Archbishop Courtney, is preserved an instrument dated 16th January, 1390, containing an account of the exchange of this advowson for the manor of Waddon, made by that archbishop and Richard Dunton, Prior of St. Saviour's, Bermondsey. Having obtained the King's licence, dated 13th December, 14 Richard II., together with the pope's bull, the case was argued on both sides in the church of Croydon, before Robert Bragbooke, Bishop of London, the pope's delegate, who confirmed the exchange. It was also agreed, by an indenture made between the above parties, that the collation and patronage of the vicarage should continue in the archbishop and his successors; and that, upon a vacancy, the archbishop or his successors should propose two fit persons to the prior and convent, who should nominate and appoint to the said vicarage.

On the 16th February, 1417, we find Archbishop Chichele issuing a commission, requiring John, Bishop of Sorron, to reconcile this church and church-yard, which had been then lately polluted by blood—The cause and manner of this bloodshed is untold; and the country being at that time internally at peace, we are led to suppose that it arose from some popular affray*.

After the dissolution of the convent of St. Saviour's in 1538, the advowson of the vicarage returned to the archbishop; and the great tithes, with the rectory manor and middle chancel, as part of the possessions of that convent, were granted to Thomas Walsingham, Esq., of Chislehurst, and Robert Moyse, Esq., of Bansted. In the will of Sir N. Heron, of Agecomb, proved in 1567, he devises his *parsonage* of Croydon to his wife Mary, for life; remainder to his son and heir Poynings† and his issue; remainder to his sons, John, William, and Henry, in succession. In the 32 Eliz., it was the property of John, Lord St. John of Bletsoe‡; and, in 1659, it had again reverted to the Walsingham family; for, it appears, that in that year, Sir Thomas Walsingham conveyed the estate to James Walsingham; and a James Walsingham, by will, dated in 1727, devised the same to his sister, Lady Elizabeth Os-

* Reg. Chichele, fol. 331 a.

† Captain Poynings Heron resided at Croydon, and held the command of 375 foot during the expected landing of the Spanish armada forces. Vide abstract of returns of able-bodied men, captains, arms, &c., April, 1588. (Harl. MSS. No. 168, p. 166).

‡ Terrier of Lands in Surry, Brit. Mus. No. 4705, Ayscough's Cat.

borne, for life; but made no ulterior bequest of it. He died in 1728, without issue, leaving three coheirs, Lady E. Osborne, Anthony Viscount Montague, and Mrs. Villiers. Lady Elizabeth, at her death, devised her third to Henry Boyle, Esq., afterwards Walsingham; from whom it descended to the Hon. Robert Boyle Walsingham; who conveyed it in 1770 to Anthony Joseph Viscount Montague, who, inheriting his father's third, and purchasing that of Mrs. Villiers, became possessed of the whole. He died in 1789; and the trustees under his will sold part of the tithes to various land-holders, and conveyed the remainder to George Samuel Viscount Montague; who was unfortunately drowned at Schaffhausen on the Rhine in 1793. In this year he had disposed of the remaining tithes, and conveyed the manor with the middle chancel to Robert Harris, Esq., who died in 1807. The trustees under whose will sold the same to Alexander Caldcleugh, Esq., of Broad Green; whose son, Alexander Caldcleugh, Esq., is the present possessor.

There were formerly two chauntries in this church, one dedicated to St. Mary, the other to St. Nicholas; that of St. Mary was founded some time previous to the year 1402, by Sir Reginald de Cobham, Lord Cobham*, of Sterborough

* Reginald de Cobham, Lord Cobham, of Sterborough Castle, Lingfield, was born about 1300; and, having distinguished himself

Castle, Surrey. The incumbent was required to pray for the repose of the souls of the said Lord Cobham, his wife Joan, daughter of Thomas second Lord Berkley, his children, and of all faithful Christians. The presentation of the chauntry priest was vested in twelve of the principal inhabitants of the town.

The other, dedicated to St. Nicholas, was founded for the repose of the souls of John Stafford, Bishop of Bath and Wells, and of William Oliver, Vicar of Croydon. This chauntry must have been founded before 1443, as in that year Bishop Stafford was translated to the see of Canterbury. The patronage of this chauntry appears to have been in the Weldon family, and was first enjoyed by Richard Weldon, Esq., who presented in right of his wife.

The interior of the church consists of two aisles, a spacious nave, and chancel; and measures, exclusive of the tower, about 130 feet in length and 74 feet in width. The nave is sepa-

in the French wars, was created a banneret by Edward III., who deputed him a Commissioner for the management of several treaties with France, particularly that of Bretigny, 1360, when he renounced his title to the kingdom. He held a principal command in the English army at the battles of Cressy and Poitiers, and was raised to the peerage in 6 Edw. III. He married Joan, daughter of Thomas Lord Berkley; and died of the plague, October 5th, 1362, leaving Reginald his son and heir.

rated from the aisles by handsome clustered columns, which support pointed arches, and was formerly divided from the chancel by twelve wooden screens of curious workmanship. These screens were taken away about the year 1817, to make room for the children of the school of industry; at which time the pulpit was removed to its present situation, and the monuments cleaned and restored. During the execution of these works, a doorway, leading to a circular staircase in the south-east column of the nave, was discovered in the chauntry of St. Nicholas.

" In the rebellion," says Aubrey, " one Bleese was hired, for half a crown *per* day, to break the painted glass windows, which were formerly fine*."

On the left of the altar stands the font, of an octagonal form, supposed to be coeval with the church; it is of white marble, with quatrefoil panels on its sides, filled with grotesque heads and roses. Before the altar stands a brass eagle, with extended wings.

The organ, a remarkably fine one, erected in 1794, was the work of Avery, who always considered it his *chef d'œuvre*.

On one of the brass chandeliers is this inscription:—

This Branch erected in the year 1717. John Bowles & Luke Bird, Churchwardens.

* Antiquities of Surrey, Vol. 2, p. 30.

The following inscriptions are to be found on the exterior of the church. On the east end of the middle chancel:—

THIS CHANCEL WAS REPAIRED & BEAUTIFIED BY ALEX[R.] CALDCLEUGH, ESQ[R.], IN THE YEAR 1808.

On the east end of St. Mary's chauntry:—

THIS CHANCEL END REPAIRED 1817, KNEVIT LEPPINGWELL, THOMAS HEWSON, CHURCHWARDENS.

On the east end of St. Nicholas' chauntry:—

1815.

THIS CHANCEL END REPAIRED, JAMES ROGERS & FRANCIS SIMMONDS, CHURCHWARDENS.

On the north entrance, where has lately been placed a neat porch, bearing, in the spandrels of the arch, the arms of his present grace, (*Az.* an eagle displayed *erminois*, charged on the breast with a cross fleury *gu.*), and a mitre.

Erected Anno Domini 1829. William Inkpen & John Brooker, Churchwardens.

In the interior, on the west gallery, is inscribed—

THIS CHURCH REPAIRED AND BEAUTIFIED, ANNO DOMINI 1823, W[M.] JOHNSON, W[M.] BLAKE, CHURCHWARDENS.

On the 25th December, 1639, a violent storm of wind blew down one of the pinnacles of the steeple, which fell upon the roof, and did great damage*; and, in 1744, the church was considerably damaged by lightning.

* History and Troubles of Archbishop Laud, p. 57.

On the 11th March, 1734-5, between two and three o'clock in the afternoon, a fire was discovered in the roof of the middle chancel, which was supposed to have been caused by some embers carelessly left there by the plumbers. It was soon extinguished, and the damage done did not exceed fifty pounds*.

The following bishops were consecrated in this church:—

1534. April 19.—By Archbishop Cranmer, Thomas Goodrich, D.D., Bishop of Ely†, and John Capon, alias Salcot, L.L.D., late Abbot of Hyde, Bishop of Bangor.

1541. September 25.—By the same archbishop, John Wakeman, last Abbot of Tewkesbury and first Bishop of Gloucester‡.

1551. August 30.—By the same archbishop, John Scory, D.D., Bishop of Rochester§, and Miles Coverdale, D.D., Bishop of Exeter.

1591. August 29.—By Archbishop Whitgift, Gervase Babington, D.D., Bishop of Llandaff‖.

1612. September 20.—By Archbishop Abbot, assisted by John (King) Bishop of London, Richard (Neile) Bishop of Lichfield and Coventry, and John (Buckeridge) Bishop of Rochester, Miles Smith, D.D., Bishop of Gloucester¶.

The vicarage-house adjoining the church-yard, was rebuilt by Archbishop Wake, in 1730, on the instigation of his lady, at a cost of about 700*l.*** A house had been appropriated to the vicar so early as the reign of Edward III.††

* Parish Register.

† Reg. Cranmer, fol. 162 a.

‡ Ibid. fol. 171 a.

§ Ibid. fol. 334 a.

‖ Strype's Life of Whitgift, p. 382.

¶ Parish Register.

** Mills' Essay on Generosity.

†† Pat. 5 Ed. 3, Pt. 1, m. 28.

RECTORS OF CROYDON.

Ægidius de Audenardo was rector in 1282* and 1295†. He was canon of the church of St. Mary, Dover, and prebend of Pesmere‡, also rector of Cherryng, which he resigned May 4, 1284§.

John Maunsel was rector in 1309||, and in 1310¶.

Richard Aungerville, al' de Bury, cl', presentat. per regem ad eccl' de Croydon, archiepatu vac', 30th November, 1 Ed. III.** This divine, afterwards Bishop of Durham, and author of the "Philobiblon," was the son of Sir Richard Aungerville, and derived his usual designation of De Bury from St. Edmund's Bury, Suffolk, where he was born, in the year 1287. He was educated at Oxford, became tutor to Prince Edward, afterwards Edward III., and receiver of the prince's revenues in Wales. On the accession of Edward III. he was appointed cofferer to his Majesty, treasurer of the wardrobe, and clerk of the privy seal, for five years; within which time he twice, as legate, visited the pontiff John, who nominated him

* Reg. Peckham, fol. 146 a.

† Ibid. fol. 97 b.

‡ Ibid. fol. 36 b.

§ Ibid. fol. 207 a.

|| Ibid. fol. 52 a.

¶ Reg. Winchelsey, fol. 19 b.

** Pat. 1 Ed. 3.

chaplain to his principal chapel, and who gave him a bull, preferring him to the first vacant see in England. The King also, in the first six years of his reign, presented him to two rectories, (including that of Croydon), six prebendal stalls, the archdeaconries of Salisbury and Northampton, the canonry of Weston, and the deanery of Wells*. On the death of Beaumont, Bishop of Durham, the prior and chapter elected Robert de Graystanes to the see, who was consecrated by the archbishop of that province, 14th November, 1333; but the King refusing his consent, Graystanes was deposed, and the Pope conferred the bishopric on De Bury, who was consecrated 19th December this year. In 1334 he was appointed lord high chancellor and high treasurer of England, and died at Auckland, 14th April, 1345.

John de Tounford was rector in 1348†.

William de Leghton, collated by Archbishop Islip, 12th January, 1351‡.

William de Wittleseye, collated by his uncle, Archbishop Islip, 12th April, 1352§. He afterwards became doctor of canon law at Oxford, and was preferred by his uncle to the office of vicar general, then to the deanery of arches, the archdeaconry of Huntingdon, the bishoprics of Ro-

* Tanner.

† Reg. Courtney, fol. 176 b.

‡ Reg. Islip, fol. 259 a.

§ Ibid. fol. 263 b.

chester and Worcester, and at last became Archbishop of Canterbury. He died 1374, having some time before exchanged this rectory for that of Clive, in the deanery of Shoreham, with

Adam de Honton, LL.D., afterwards Bishop of St. David's, collated 3rd May, 1359*. He was canon of St. David's, consecrated bishop in 1361, appointed chancellor of England, January 11, 1377, which office he quitted 1379, and died 13th February, 1389. He built St. Mary's college, near his cathedral, which he endowed with 100*l.* *per annum.*

Adam de Robelyn was rector in 1363†. He exchanged this rectory for the prebend of Ruyll, in the collegiate church of Abergwilly, with

William Bourbrigg, who was admitted 8th June, 1363‡.

John Quernby was rector in 1364§. He exchanged this rectory for the prebend of Woodburgh, in the collegiate church of Southwell, York, with

John Godewyke, admitted 28th March, 1365‖.

John Godewyke, LL.D., presented on the 6th November, 1370, by Edward III., who became patron, on the temporalties of the vacant arch-

* Reg. Islip, fol. 282 b.
† Ibid. fol. 301 a.
‡ Ibid.
§ Ibid. fol. 306 b.
‖ Ibid.

bishopric being in his hands*. He was the last rector of this church.

Vicars of Croydon.

Henry de la Rye, presented to this vicarage by Ægidius de Andenardo, rector of the same, 4th August, 1289†.

Thomas de Sevenoke is mentioned as vicar in 1309‡.

Thomas de Maydenestan, presented by John Maunsel, rector, in May, 1309‡.

John de Horstede is stated to be vicar in 1348§.

John de Stanesfelde, who was appointed Dean of Croydon|| by a commission from Archbishop Islip, dated at Lambeth, 11th February, 1349¶. He exchanged this vicarage for the rectory of West Wickham, with

Richard atte Lich', presented by William de Wittleseye, rector, 7th June, 1356**.

* Wittleseye, fol. 82 b.

† Reg. Peckham, fol. 40 a.

‡ Reg. Winchelsey, fol. 82 b.

§ Reg. Courtney, fol. 176 b.

|| The deanery of Croydon was composed of the churches of Croydon, East Horsley, Merstham, Wimbledon, Barnes, Burstow, Charlwood, Newington, and Cheyham. Croydon is now in the deanery of Ewell.

¶ Reg. Islip, fol. 10 a.

** Ibid. fol. 271 b.

John de Hamaldon, presented by Adam de Honton, LL.D., rector, 3rd December, 1361 *.

Robert Okele, presented by John Godewyke, rector, May, 1373 †.

John Lane, upon whose resignation

John Brown was presented ‡.

William Daper was rector in 1402 §. He exchanged this vicarage for the rectory of Throckyng in the diocese of Lincoln, with

Richard Bondon, presented 7th August, 1402, by the convent and prior of St. Saviour's, Bermondsey §. He exchanged this vicarage for the wardenship of St. Mary Magdalen, with the parish of Kingston, with

John Scarburgh, who was presented by the same patrons, 18th December, 1405 ||.

John Aldenham, alias Causton, presented by the same patrons 20th January, 1408 ¶.

The vicarage became vacant before 23rd November, 1420; but in what manner we are not told, a blank being left in Archbishop Chicheley's register, (on the side of which is written "Institutio Vicarii de Croydon"), where

* Reg. Islip, fol, 243 b.

† Reg. Wittlesey.

‡ Reg. Courtney.

§ Reg. Arundel, Part 1, fol. 284 a.

|| Ibid. fol. 305 b.

¶ Ibid. Part 2, fol. 52 a.

the name of his successor should have been inserted*.

William Oliver*.

John Langton; upon whose death†

Henry Carpenter, LL.B. was presented by the prior and convent of St. Saviour's, Bermondsey, 30th October, 1487†.

William Shaldoo, presented by the same patrons, 3rd December, 1487‡.

Roland Phillips, D.D., collated June 4, 1497§, by Archbishop Morton, with the unanimous consent of the prior and convent of St. Saviour's, Bermondsey. He was a canon of St. Paul's, prebend of Bryghtling in the collegiate church of Hastings, 9th June, 1507||; rector of St. Margaret Pattens, which he resigned in 1515; rector of St. Michael's, Cornhill, 14th August, 1517¶; prebend of Measdon, in St. Paul's, 28th November, 1517¶; rector of Mestham, Surrey, 1520||; and warden of Merton College. This vicar, preaching at St. Paul's against printing, then lately introduced into England, uttered the following singular passage: "We" (the Catholics)

* Reg. Chichele, Part 1, fol. 121 b. † Bourchier, fol. 97 b.

‡ Reg. Morton, Dene, Bourchier, Courtney, fol. 133 a.

§ Reg. Morton, fol. 163 a. || Reg. Warham.

¶ Reg. Fitz-James at Stokesly.

"must root out printing, or printing will root out us*."

Of this celebrated preacher, of whom so little is known, we subjoin the following notices. It is to be regretted that no memoir, however brief, of this man, apparently so notorious in his time, should exist.

Roland Phillips.

"This king (Henry VIII.) came to the title of 'Defender of the Faith' (A.D. 1528) when Luther had uttered the abominations against the Pope and his clergy, and divers books were come into England, our cardinal (Wolsey) to find a remedy for it, sent to Rome for the title of 'Defender of the Faith.' After, the vicar of Croydon preached that the king would not lose it for all London and twenty miles about it†."

"And even as there was much ado amongst them of the Common House, about their agreement to the subsidie, so was there as harde holde for a whyle amongst them of the clergie in the Convocation House: namelye, Richard Byshoppe of Winchester, and John Byshoppe of Rochester, held sore agaynst it; but, most of al, Sir Rowlande

* Fox. † Acts of the Church, 1614, p. 169.

Philips, vicar of Croydon, and one of the canons of Paules, being reputed a *notable preacher* in those dayes, spake most against that payment. But the cardinall taking him aside, so handled the matter with him, that he came no more into the house, willingly absenting himselfe, to his great infamie and losse of that estimation which men had of his innocencie. Thus, the Bellweather, giving over his holde, the other yielded, and so was granted the halfe of all their spirituall revenues for one year, to be paid in five yeares following, that the burthen might y[e] more easily be borne*."

"Yet, because he (Sir T. More) would not blame anie man's conscience therein, he was commanded to walke into the garden a while; and presently all the clergie men, some bishops, manie doctours and priests were called in, who all took it, except Bishop Fisher and one Doctour Wilson, without anie scruple, stoppe, or stay; and the Vicar of Croydon, saith Sir Thomas, called for a cuppe of biere of the butterie barre, *quia erat notus pontifici*, and he drunke *valde familiariter*†."

"The Vicar of Croydon, under the archbishop's

* Holinshed's Chron. p. 1524. See also Stowe.

† More's Life of Sir T. More, p. 222.

nose, had been guilty of certain misdemeanours, which, I suppose, were speaking or preaching to the disparagement of the king's supremacy, and in favour of the Pope. Now, before he went into the country, and having as yet divers bishops and learned men with him at Lambeth, he thought it advisable to call this man before them at this time; but before he would do it, he thought it best to consult with Crumwell, and take his advice whether he should now do it, and before these bishops or not*."

"Besides these, were three that supplicated that were not admitted this year (1515), of whom Rob. Schowldham, before mentioned, was one, and Rowl. Phillips, M.A., ***an eminent preacher of his time,*** afterwards warden of Mirt College, and another†."

"January 2, 1522. Rowland Philips, M. of A., supplicated for the degrees of bach. and doctor of divinity, and was, as it seems, admitted. Soon after, by the power of the Archbishop of Canterbury, he was thrust in warden of Merton College. He was now vicar of Croydon in Surry, one of the canons of Pauls, ***a famous and notable preacher, and a forward man in the convocation of the clergy,*** an. 1523, in acting and speaking

* Strype's Life of Cranmer, p. 79.

† Wood's Ath. Ox. by Bliss, Vol. 2, p. 41.

much against the payment of a subsidy to the king*."

"He (Ruthall, Bishop of Durham) paid his last debt to nature, at Durham Palace, near London, on Wednesday the fourth of Feb. in fifteen hundred twenty and two, and was buried in the chapel of S. John Baptist, joyning to the abbey church of S. Peter in Westminster; at which time Dr. Rowl. Phillips, Vicar of Croydon, *a great and a renowned clerk,* preached an excellent sermon†."

He also attended the funeral of Abbot Islip, at Westminster, in 1532, and preached his funeral sermon‡.

Peter Burough, M.A., collated on the resignation of Roland Phillips, by Archbishop Cranmer, 9th May, 1538§, *pleno jure;* on which day the archbishop issued a decree to John Cocke, LL.D., his vicar-general, to assign a pension of 12*l.* *per annum* from the profits of the vicarage to the said Roland Phillips, for life, on account of his great age.

John Gibbes, B.D., collated by the same

* Wood's Ath. Ox. Vol. 2, p. 61. † Ibid. p. 723.

‡ Widmore's Hist. of West. Abbey, Appendix, 10.

§ Reg. Cranmer, fol. 364 b.

archbishop, 12th April, 1542*. Of this preferment he was, however, deprived for refusing to pay his tithes to the King; when he was succeeded by

David Kemp, collated by the same archbishop 31st May, 1550†; on whose resignation

William Cooke was collated by the same archbishop, 13th September, 1553‡; upon whose death

Richard Fynche was collated by Archbishop Parker, 23rd April, 1560§; upon whose death

Samuel Fynche was collated by Archbishop Grindall, 26th May, 1581||.

Samuel Fynche, at the presentation of the king by lapse, 28th February, 1603; upon whose death¶

Henry Rigge, M.A., was collated by Archbishop Abbot, 20th September, 1616**.

Samuel Bernard, B.D., justice of the peace for the county, was collated, on the resignation of Henry Rigge, 10th August, 1624††. He was of Magdalen College, Oxford, and proceeded D.D. March 15, 1638. He was displaced by the committee for plundered ministers, in Feb-

* Reg. Cranmer, fol. 364 a.
† Ibid. fol. 411 a.
‡ Ibid. fol 424 a.
§ Reg. Parker, Part 1, fol. 342 b.
|| Reg. Grindall, fol. 551 b.
¶ Reg. Whitgift, Part 3, fol. 278 b.
** Reg. Abbot, Part 1, fol. 420 a.
†† Id. Part 2, fol. 337.

ruary, 1643, "for errors in doctrine, superstition in practice, and malignancy*." He was afterwards rector of Farley, Surrey, and author of a funeral sermon on Ezek. xxiv. 16, Lond. 1652. Antony à Wood says—"he was a Berkshire man born, and had, in his younger days, been accounted a good Greek and Latin poet." His successor was

Thomas Buckner, D.D.†, who was succeeded by

Samuel Otes, M.A., who lies buried in the north chancel. He died in 1645.

Francis Peck. In 1646, it was ordered by the committee that 50*l* *per annum* should be paid to Francis Peck, out of the impropriate rectory of East Meon in Hampshire, as an augmentation of the vicarage of Croydon. This money having never been received, the same sum was voted, 20th September, 1684, to his successor,

Edward Corbett, M.A., out of the sequestered rectory of Camberwell‡.

Jonathan Westwood, appointed by Sir William Brereton, baronet, who was ordered by the com-

* Walker's List of Ejected Clergy, p. 210. He died August 5, 1657, and was buried at Farley, where he is described on his tomb as "Pastor fidus, vir nullo fœdere fœdatus."

† "Samuel Barnard being displaced in 1643, Thomas Buckner, D.D., was appointed, but died in 1644."—Rawlinson's MS. notes on Aubrey.

‡ Proceedings of the Committee, Bodleian Library.

mittee for reformation of the universities, to provide a minister to serve the cure of the church of Croydon; for whose stipend they assigned to him 50*l.* *per annum.* I have not been able to discover the date of the appointment of this minister, nor the time of his cession (by death or otherwise); but certain it is that he was in receipt of the above stipend from the 31st May, 1654, to the 9th June, 1657*. He was probably succeeded by—

William Cleiver, D.D., collated by Archbishop Juxon in 1660†, at the recommendation of Charles II., who had been imposed upon with regard to his character‡. This divine was notorious for his singular love of litigation, unparalleled extortions, and criminal and disgraceful conduct, which eventually caused his ejectment from this benefice in 1684§. He was a great persecutor of the royalists during the Commonwealth; and enjoyed the sequestered living of Ashton, Northamptonshire, to which he was appointed in 1645||, being at that time, according to Walker, scarce eighteen, "of a very ill life, and very troublesome to his neighbours¶." He died in March 1702; and

* Vide Appendix, No. V.

† Parish Register.

‡ Vide Appendix, p. 365.

§ Vide "Case of the Inhabitants of Croydon," Appendix, No. 11.

|| Bridge's Northamptonshire, Vol. 1, p. 284.

¶ Walker's List of the Ejected Clergy, Part 2, p. 402.

was buried on the 12th of that month at St. Bride's, Fleet Street; in the register of which church he is styled "Parson of Croydon."

The following anecdote of this vicar is to be found in Captain Smith's "Lives of Highwaymen:—" O'Bryan, meeting with Dr. Cleiver, the parson of Croydon, *try'd once and burnt in the hand at the Old Bailey for stealing a silver cup,* coming along the road from Acton, he demanded his money; but the reverend doctor having not a farthing about him, O'Bryan was for taking his gown. At this our divine was much dissatisfied; but, perceiving the enemy would plunder him, quoth he, 'Pray, Sir, let me have a chance for my gown;' so, pulling a pack of cards out of his pocket, he farther said—'We'll have, if you please, one game of all-fours for it, and if you win it, take it and wear it.' This challenge was readily accepted by the foot-pad, but being more cunning than his antagonist at slipping and palming the cards, he won the game, and the doctor went contentedly home without his canonicles*.

Henry Hughes, M.A., collated by Archbishop Sancroft, 26th June, 1684†; on whose resignation

John Cæsar, M.A., was collated by the same archbishop, 18th January, 1688‡. There is a sermon preached by this vicar, 10th March, 1707, at the Croydon Assizes, printed 1708, where he

* Vol. 1, p. 257.

† Reg. Sancroft, fol. 404 b.

‡ Ibid. 425 b

is styled Vicar of Croydon and Chaplain to Scroop Earl of Bridgewater. On his death,

Andrew Trebeck, B.D., was collated by Archbishop Wake, 28th April, 1720. Upon whose resignation,

Nathaniel Collier, M.A., was collated by the same archbishop, 29th November, 1727. Upon whose death

John Vade, M.A., was collated by Archbishop Herring, in January, 1755. He was vicar of St. Nicholas, Rochester.

East Apthorp, D.D., collated by Archbishop Secker, on the death of John Vade, in June, 1765*. He was rector of St. Mary-le-Bow, London, prebend of Finsbury, and author of "Letters on the Prevalence of Christianity." Upon whose resignation

John Ireland, D.D., was collated by Archbishop Moore, 15th July, 1793. This divine, who is at present Dean of Westminster, is author of "Five Discourses, containing certain arguments for and against the reception of Christianity by the Ancient Jews and Greeks, 1796." Upon whose resignation

John Cutts Lockwood, M.A., was collated by Archbishop Sutton, 30th March, 1816. He was

* Dr. Apthorp died 17th April, 1816, at Cambridge, where he had formerly been Fellow of Jesus' College.

also rector of Coulsdon, Surrey. Upon whose death

Henry Lindsay, M.A., perpetual curate of Wimbledon, Surrey, the present vicar, was collated by the Primate, 4th November, 1830. He is author of "Practical Lectures on the Historical Books of the Old Testament."

INCUMBENTS OF ST. MARY'S CHAUNTRY.

John Parke occurs in 1402*. Upon whose resignation

Clement Ecclestone was presented 7th August, 1402†. He exchanged the chauntry for the rectory of Depedon, Winchester, with

Stephen Alchon, admitted 19th September, 1409‡. Upon whose resignation

Robert Peterburgh was presented, 27th February, 1420§. Upon whose death

Thomas Barfote was presented, 3rd March, 1430‖. Upon whose death

William Kyng was presented, 5th March, 1458¶. Upon whose death

Thomas Thomlynson** was presented, 12th June, 1476. Upon whose resignation

* Reg. Arundel, Part 1, fol. 284 a.

† Ibid.

‡ Ibid. Part 2, fol. 54 b.

§ Reg. Chichele, Part 1, fol. 71 b.

‖ Ibid. fol. 188 b.

¶ Reg. Bourchier, fol. 74 a.

** Ibid. fol. 113 b.

John Knowdyson* was presented 17th October, 1499.

Edward Jenyns† occurs in 1505. Upon whose resignation

Andrew Corphell‡ was presented, 23rd October, 1505. Upon whose death

John Comporte§ was presented, 4th September, 1538. He was the last incumbent, and had a pension of *6l.* 13*s.* 4*d.* granted him for life, at the dissolution of this chauntry, 1 Edward VI.¶; at which time its revenue amounted to 16*l.* 1*s.* 2*d.*

INCUMBENTS OF ST. NICHOLAS' CHAUNTRY.

Henry Foxwyst**.

Robert Smyth††, collated by Archbishop Stafford, 30th June, 1450.

John Gosse‡‡ occurs in 1454. He exchanged this chauntry for the rectory of Grendone, Lincoln, with

* Reg. Morton, Dene, &c., fol. 168 a.
‡ Ibid.
† Reg. Warham, fol. 365 b.
§ Reg. Cran. fol. 365 b.
¶ By the statute of the 1 Edward VI. c. 14, all chauntries were abolished.
** Reg. Chichele, Part 1, fol. 233 b.
†† Reg. Stafford and Kemp, fol. 105 a.
‡‡ Reg. Bourchier, fol. 59 a.

John Meyskyn*, presented 7th November, 1454, by Richard Weldon, Esq., patron and founder, in right of Elizabeth his wife.

William Walton† occurs in 1472. Upon whose death

William Spynke‡ was presented 17th January, 1472, by Richard Weldon, Esq.

Nicholas Brooke§, presented by the same patron, 13th August, 1474, on the resignation of William Spynke. Upon whose resignation

Robert Dady|| was presented by the same patron, 16th March, 1479. Upon whose death

Robert Hollere¶, M.A., was presented by Elizabeth, widow of Richard Weldon, Esq., 9th February, 1487. Upon whose death

Thomas Greene** was presented by the same patron, 10th October, 1591. Upon whose resignation

John Maynell†† was presented by Robert Weldon, Esq., 17th June, 1490.

Thomas Sparke‡‡ occurs in 1504. Upon whose resignation

Henry Molle§§ was collated by Archbishop

* Reg. Bourchier, fol. 59 a.

† Ibid. fol. 107 a.

‡ Ibid.

§ Ibid. fol. 111 a.

|| Ibid. fol. 124 b.

¶ Reg. Morton, Dene, &c., fol. 133 a.

** Ibid. fol. 153 b.

†† Ibid. fol. 168 a.

‡‡ Reg. Warham, fol. 322 a.

§§ Ibid.

Warham, 19th December, 1504; to whom the right of presentation devolved, by the neglect of Ellen, widow of Robert Weldon, Esq., and of the vicar and churchwardens of Croydon, who omitted to present within the time limited by the founder of the said chauntry. Upon his death,

Richard Parrer* was presented by Ellen Weldon, widow, and Hugh Weldon, Esq., 31st January, 1508. Upon whose resignation

Henry Marshall† was presented by the same patrons, 2nd June, 1509. Upon whose death

William Shanke‡ was collated by Archbishop Warham; to whom the right had devolved, 19th October, 1521. Upon whose death

Nicholas Sommer§ was presented by Hugh Weldon, Esq., 11th May, 1531. He was the last incumbent, and had a like grant of 6*l.* 13*s.* 4*d.* for life. The income of this chauntry was 14*l.* 14*s.* 6*d.*

Croydon Church Register.

The first Croydon register commences in 1538, when Cromwell, vicar-general, issued an order for parish registers to be kept throughout the kingdom. The first and second volumes are bound in russia leather, at the expense, it appears, of

* Reg. Warham, fol. 334 a.

† Ibid. fol. 335 a.

‡ Reg. Warham, fol. 394 b.

§ Ibid. fol. 404 b.

Sir Isaac Heard, Knt., Garter-King-at-Arms. From these documents we learn, that, from July 20, 1603, to April 16, 1604, one hundred and fifty-eight persons died of the plague at Croydon. In 1625, the number amounted to seventy-six; in the following year to twenty-four; and in 1631, to seventy-four. From July 27, 1665, to March 22, 1666, one hundred and forty-one persons died of this pestilence.

The following miscellaneous items are arranged according to their respective dates:—

June 10, 1552. Alexander Barckley, sepult*.

Anno dni 1560. Syr Wyllm Coke, clerke, vicar of Croydon, was buryed the xxvij day of Marche.

1563. January. Mr. Wyllm Heron, justice, was buryed the x day of January.

1568. Syr Nicholas Heron, Knight, deceased the fyrst day of September, and was buried the v day of the same month.

1578. Lady Mary Heron [widow of Sir Nicholas] was buryed the xx day of Aprill, and her funerall was made the xxiiij day of Aprill.

1578. This Candlemas was the great snowe.

1581. Richard Ffinche, clerke, vycar of the paroyche churche of Croydon, was buryed the ixth day of Aprill, anno dni 1581mo, regni Eliz. 23tio.

Edmunde Grindall, L. Archbishop of Canterburie, deceased the vj day of Julye, and was buried the fyrste day of Auguste, anno dni 1583, and anno regni Elizabethæ, 25.

In his will he ordered his body to be buried "in the choir of the parish church of Croydon, without any solemn herse or funeral pomp."

* For an account of this celebrated writer, vide ante, pp. 21 and 22.

1584. Bonaventure Ryder, travelynge between Wonswthe and Croydon, was found dead in Waddon mill, upon the xxv day of Julye, and was buried the iiij day of August abovesayd.

Elizabeth, the daughter of John Kynge and Clemence, (wyfe of Samuell Ffynche [primus], vycar, by the space of vij yeares), mother of v children at severall byrthes, of the age of xxj yeares; deceased the xvijth day of November, and was buryed the xviijth, anno dni 1589.

Memoranda.—That whereas Samuell Ffynche, vicar of Croydon, lycensed Clemence Kinge, the wyfe of John Kynge, brewer, to eate fleshe in the time of Lente, by reason of her sicknesse, wch lycence beareth date the xxixth day of Ffebruary; and further, that she the sayde Clemence doth as yet contynue sicke, and hath not recovered her health; Knowe ye therfor, that the sayde lycence continueth still in force, and for the more efficacie therof, ys here registered accordinge to the statute, in the psence of Thomas Mosar, churchwarden of the said parishe of Croydon, the vijth day of Marche, in the xxxviij year of the Queene's mat[ts] moste gracious reigne, and for the registeringe therof ther is paid unto the curate iv *d*.

A like licence was granted by the same vicar to Susan Weller of Croydon, on the 25th March, 1598.

John Whitgifte, Archbishop of Canterburie, deceased at Lambith on Wednesday, at viiij of the clocke in the eveninge, beinge the laste day of February, and was brought the day followinge in the eveninge to Croydon, and was buried the morninge followinge, by two of the clocke, in the chappell where his pore people doe usuallie sitte; his ffunerall was kepte at Croydon, the xxvijth day of Marche followinge, anno dni 1604, anno. regni dni nri Regis Jacobi Secundo.

His funeral was performed with great solemnity, the Earl of Worcester and Lord Zouch bearing the pall, and Dr. Babyngton, Bishop of Worcester, preaching the funeral sermon from the following text:—" But Jehoiada waxed old, and was full of days, and died. An hundred and thirty years old was he when he died. And they buried him in the city of David, with the kings, because he had done good in Israel, and towards God and his house."

December, 1607. The greatest ffrost began ye ixth day of this month. Ended on Candlemas-eve.

1608. Mychaell Murgatrode, Esquire, deceased at London the x day, and was buried at Croydon on the xii day of Aprill, anno dni 1608.

This gentleman was fellow of Jesus College, Cambridge, and secretary to Archbishop Whitgift*.

Ffrancis Tyrrell, citizen and marchante of London, was buried the first day of September, 1609, and his ffunerall was kept at London the xiijth of the same month. He gave two hundred poundes to the parishioners of Croydon, to builde them a newe market house, and ffortie pounds to repaire our churche, and ffortie shillings a yeare to our pore of Croydon, for xiij years, withe manie other good and greate legacyes to the citie of London.

Charles Howarde, sonne unto the Righte Honourable Charles Earle of Nottingham, borne the xxvth daye of December, anni dni 1610, was christened the xxiijd daye of January followinge.

This Charles Howard was the only son of the Lord Admiral, by his second wife, Lady Margaret Stewart, daughter of James Earl of Moray, and became third and last Earl of Nottingham. He married Arabella, daughter of Edward Smith, Esq., and died 26th April, 1681.

October, 1616. Samuel Finch [secundus], vicar of Croydon, was buried the 15th day.

* Among the Harl. MSS. (No. 67) is a volume labeled "Murgetrodi et Bell. Orat. et Epist." The title of the oration is—"De Græcarum Disciplinarum Laudibus Oratio, quam apud Jesuanos auditores suos habuit;" and inscribed Mich. Murg.

George Abbot, Lord Archbishop of Cant., deceased at Croydon upon the fourth day of August, 1633. His funerall was with great solemnity kept in the church here, upon the third day of Septemb. following, and the next day his corpse was convaide to Guilford, and there buryed, according to his will.

1643. May 12. Sir Hugh Wirrall, knight, was buried.

1649. March 29. My Lady Scudamore buried.

Lady Scudamore was aunt to the patriot Hampden, and to Edmund Waller, the poet.

1650. February. S[r] John Tonstall was buryed.

1651. February. My lady Tonstall was buryed.

1667. John Davenante, Cytizen of London, was buried the xxviijth day of October.

This John Davenant, of the same family as the celebrated Sir William D'Avenant, was father to John Davenant, D.D., Bishop of Salisbury.

1677. November 16. Gelbert Sheldon, laite Archbishop of Canterbury, buryed.

Archbishop Sheldon was buried in a private manner, according to his express order.

Dr. William Wake, Archbishop of Canterbury, died at his palace at Lambeth, Jan. 24, 1736, and was brought to Croydon, and buried Feb. 9; and his lady, which was buried at Lambeth the April 1731, was taken up and brought to Croydon the next day, and put in the vault with him.

Dr. John Potter, Archbishop of Canterbury, was buried Oct. 27th, 1747.

1749. August 30. James Cooper, a highwayman, was executed on a gibet in Smithden Bottom, and their hanged in chains, for murdering and robing of Robert Saxby, groom to John How, Esqr. of Barrowgreen in the parish of Oxteed in Surry, on the 17th of March, 1749, near Crome Hurste.

The following lines, preserved by Ducarel, who had not met with the name of the murdered man, or the circumstances, were formerly on a rail in the church-yard:—

Thou shalt do no murder, nor shalt thou steal,
Are the commands Jehovah did reveal;
But thou, O wretch! who, without fear or dread
Of thy tremendous Maker, shot me dead,
Amidst my strength and sin—but, Lord, forgive,
As I, through boundless mercies, hope to live!

Dr. Thomas Herring, Archbishop of Canterbury, died at his palace at Croydon, and was buried Mar. 24, 1757.

He was buried in a very private manner, according to his own request, which expressly forbad that any monument should be raised to his memory.

Monuments and Epitaphs in the Church.

Middle Chancel.

On the south side of the altar, on a sarcophagus within an arched recess, the entablature of which is supported by Corinthian columns, lies the painted effigies of a churchman in his scarlet robes. Surmounting the entablature are three shields of arms, viz. centre shield, the arms of the see of Canterbury empaling quarterly *or* and *az.*, a cross quartered *erm.* and *or*, between four pea-hens collared and counterchanged; dexter shield, the arms of the see of York; sinister shield, the arms of the see of London, both empaling the same. Beneath his effigies are these verses:—

Grindall' doctus, prudens, gravitate verendus,
Justus, munificus, sub cruce fortis erat.
Post crucis ærumnas Christi gregis Anglia fecit
Signiferum, Christus cœlica regna dedit.

In memoria æterna erit justus.—Psal. cxii.

At the top of the monument—

Beati mortui qui in Dno moriuntur:
Requiescunt enim à laboribus suis:
Et opera illorum sequuntur illos.
Apoc. 14.

Under the above are the two following verses in juxtaposition—

Præsulis eximii ter postquam est auctus honore,
Pervigiliq greges rexit moderamine sacros:
Confectum senio durisq laboribus, ecce
Transtulit in placidam Mors exoptata quietem.

Mortua marmoreo conduntur membra sepulchro
Sed mens sancta viget, Fama perennis erit,
Nam studia et Musæ, quas magnis censibus auxit,
Grindali nomen tempus in omne ferent.

And immediately above the effigies is this inscription:—

Edmund' Grindall' Cumbriensis, Theol' D', Eruditione, Prudentia, et Gravitate clarus; Constantia, Justitia, et Pietate insignis, civibus et peregrinis charus; ab exilio (quod Evangelii causa subiit) reversus ad summum dignitatis fastigium (quasi decursu honorum) sub R. Elizabetha evectus, Ecclesiam Londinen. primum, deinde Eborac. demu. Cantuarien. rexit. Et cum jam hic nihil restaret quo altius ascenderet, e corporis vinculis liber ac beatus ad cœlum evolavit 6° Julii an. Dni 1583. Ætatis suæ 63. Hic, præter multa pietatis officia quæ vivus præstitit, moribundus maxima. bonorum suorum partem piis usibus consecravit. In Paræcia Divæ Beghæ (ubi natus est) Scholam Grammatic. splendide extrui et opimo censu ditari curavit. Magdalenensi cœtui Cantabr. (in quo puer primum Academiæ ubera suxit) discipulum adjecit, Collegio Christi (ubi adultus liris. incubuit) gratum Μνημόσυνον reliquit; Aulæ Pembrochinæ (cujus olim Socius, postea Præfectus, extitit) Ærarium & Bibliothecam auxit, Græcoq. Prælectori, uni Socio, ac duobus Discipulis, ampla stipendia assignavit. Collegium Reginæ Oxon. (in quod Cumbrienses potissimum cooptantur) nummis, libris et magnis proventibus locupletavit. Civitati Cantuar. (cui moriens præfuit) centu. libras, in hoc, ut pauperes honestis artificiis exercerentur, perpetuo servandas, atq. impendendas dedit. Residuum bonoru. Pietatis operibus dicavit. Sic vivens moriensq. Ecclīæ, Patriæ et bonis literis profuit.

To the west of Archbishop Grindall's monument, above the door, is a neat white marble tablet, inscribed—

This Tablet is erected
To the memory of the Rev[d]. Joseph Scaife, B.A., late of Queen's College, Oxford, and of Birkby in the county of Cumberland. He served for two years previous to his death the curacy of this parish, and in so exemplary a manner as to gain the esteem of many friends, who are desirous thus to record his worth and their affectionate remembrance of him.

His mortal remains are deposited at Bidborough in the county of Kent, near which place he died March 9, 1819, aged 26.

On the north side of the altar, within separate recessed arches, and flanked and divided by a Corinthian column, are the painted effigies of a man and woman kneeling before desks. Above the entablature are three shields of arms, viz. centre shield, quarterly 1 and 4 *erm.* a millrind *sa.*, 2 per pale *az.* and *gu.*, three lions ramp. *erm*, 3 *or* in a bordar engr. *sa.*, a saltire *gu.* between four pears of the last; dexter shield, *erm.* a chev. *gu.;* sinister shield, *gu.* on a fesse ingr. *arg.*, three cross pattée's *sa.*, between three waterbougets *or.*

Over the man is this inscription:—

Obiit 21 Jana 1573, æt. suæ 69.

Over the woman—

Obiit 2 Aug. 1585, æt. suæ . .

Under the man is this inscription, in capitals:—

Heare lyeth buried the Corps
Of Maister Henrie Mill
Citezen and Grocer of
London famous Cittie
Alderman and sometyme Shrive.
A man of prudent Skill,
Charitable to the Poore,
And alwaies full of pittie.
Whose Soule wee hope dothe rest in
Blise, wheare Joy dothe stil abounde
Thoughe bodie his full depe do lie
In earthe here under grounde.

Under the woman—

Elizabeth Mill his lovinge wyf
Lyeth also buried heare
Whoe sixtene Children did him beare
The blessing of the Lorde,
Eight of them sonnes, and the other 8
Weare daughters. This is cleare
A witnes sure of mutuall love
And signe of greate accorde.
Whose Sole amonge the Patryarks
In faithfull Abram's brest,
Thoughe bodie hirs be wrapt in clay,
We hope in joye dothe rest.

At the bottom of the tomb—

Ano. Dni 1575.

Immediately above the last-mentioned tomb, on a handsome white marble tablet:—

To the memory of
NICHOLSON DUNDAS ANDERSON,
Son of the late Robert Anderson, Esq., Superintendant of the Honble.
E. I. Company's Marine, at Bombay,
And a Student of the College at Addiscombe,
Who was drowned in the Croydon Canal
On the 29th day of Augst. 1818, in the 16th year of his age.

Whilst an ingenious disposition endeared him to his associates, the high character which he uniformly sustained for exemplary conduct and assiduous application held out the fairest hopes that his career in life (had its continuance been vouchsafed to him) would have proved a constant source of happiness and distinction to himself, and of comfort and satisfaction to his relations. But the ALMIGHTY hath willed it otherwise.

Reader—doubt not his wisdom or goodness even in this bitter dispensation. May he not have dealt graciously with his servant, by removing him to a better world, before his bright prospects in this life had been darkened by disappointment, or his integrity had fallen a prey to its corruptions.

To the east of the last, on a black marble tablet, supported by two Corinthian columns, and bearing the following arms—*gu.* a chev. ingr. between three estoils *arg.*, is this inscription:—

Here lyes the Body of JOHN PYNSENT, Esqr.
One of the Prothonotaries of his Majesties
Court of Comon-Pleas, who departed this
life the 29th of August 1668.

The meanest part of him is only told
In this Inscription, as this Tombe doth hold
His worser part, & both these easily may
In length of time consume and weare away;
His Virtue doth more lasting honours give,
Virtue and Virtuous souls for ever live;

This doth embaulme our dead beyonde the art
Proud Ægypt used of old; his head & heart
Prudence and pietie enricht, his hand
Justice and charity did still command;
He was the churche's and the poore man's freind;
Wealth got by Law, the Gospell taught to spend.
From hence he learnt that w[t] is sent before
Of our estates doth make us rich farr more,
Than what we leave, and therefore did hee send
Greate portions weekly; thus did he commend
His faith by workes; in heaven did treasure lay;
Which to possess his soule is cald away:
Here only is reserved his precious dust,
Untill the resurrection of the just.

"Blessed are the Dead that dye in the Lord; they rest from their Labours, and their works doe follow them."
—Rev. xiv. 13.

On the Ground.

On a brass plate, under the figure of a priest praying, in the centre of the chancel:—

Silvester Gabriel, cujus lapis hic tegit ossa,
Vera sacerdotum gloria nuper erat,
Legis nemo Sacre Divina volumina verbis
Clarius, aut vita sanctius explicuit.
Cominus ergo Deu. modo felix, eminus almis
Qem. pius in scriptis viderat ante, videt.

Anno dni Millimo v.xv. iiij. die Octobr. vita est funct.

Adjoining the last, on the south—

Here lieth Mary, the Widow of Robert Mackett, Esq., who died 22nd August, 1786, aged 83.

Adjoining the last, on the south, on a black marble ledger—

Here lie the remains of HANNAH SMITH, widow of the Rev. John Smith, Rector of Carlton in Norfolk. She died on the 6th April, 1794, in the ninetieth year of her age.

Also, the remains of PAULINA SMITH, her daughter, who died 15th January, 1813, aged 78.

To the north of Silvester Gabriel, on a black marble ledger—

MARIA DAUTRIVE,
Fil. natu min.
DANIELIS RICHARD DE WADDON,
Ob. VII Feb. MDCCLXXXVIII.
Æt. XXXI.

To the north of the last, on a black marble ledger, with arms *erm.* on a bend *gu.*, three esquires' helmets—

Here lies interred the body of Mr. JOSEPH WILLIAMS, citizen and grocer of London, who died the 15th of June, 1756, aged 57 years.

To the west of the last is a brass plate affixed to a rough grave-stone, with arms *az.* a chev. between five escallops in chief and one in base, *arg.*, with a crescent for difference, empaling cheque *arg.* and *az.*, a fesse of the last, with a crescent for difference—

Here lyeth buried the body of NICHOLAS HATCHER,
of Croydon, in the County of Surry, *Gentleman*,
Who was Captaine of a Troop of Horse under his
Most Sacred Majestie King CHARLES the First,
and Yeoman-Usher in ordinary to his Majestie
King CHARLES the Second.
Who departed this life the 29th of September,
in the year of our Lord God, 1673,
aged 69 *years.*

To the west of Silvester Gabriel, on a large black marble ledger—

ALEXANDER CALDCLEUGH, Esq.
of Broad Green,
departed this life
January 18, 1809, aged 55 years.

ELIZABETH, Daughter of the above,
and Wife of WILLIAM PLASKET, Esq.
of Old Burlington Street,
died the 24th day of November, 1832.

At the bottom of the ledger—

This is the family vault.

To the north of the last, on a black marble ledger—

DANIEL RICHARD, ESQ. of Waddon, died IX December, MDCCXCIII. Aged LXXVII.

At the head of Caldcleugh are the five following inscriptions on separate ledgers, commencing southward:—

On a black marble ledger—

Here lyeth the body of Mrs. ELIZABETH MOULTON (relict of James Moulton, Gent.), who died February 10, 1772, aged 67 years.

On a like ledger—

Here lyeth the body of JAMES MOULTON, Gent.,
Who died the 5th of October, 1761, aged 59 years,
Deservedly esteemed.
His extensive Liberality to the poor
Was an amiable example to the wealthy,
And his death a real Loss
To the aged and indigent.

On a Portland stone ledger—

Here lyeth the body of THOMAS WOOD, (youngest son of John Wood, senr., late of Woddon, by Arabella his wife), who died a bachelor Nov. the 8th, 1757, in the 47th year of his age.

To whose memory this stone is placed,
by Elizabeth (the wife of
James Moulton, Gentleman)
his sister and sole heiress.

On a Portland stone ledger—

Here lieth interred the body of JOHN WOOD of Waddon, senr., who departed this life February the 28th, 1738, in the 69th year of his age.

And also, ARABELLA WOOD, wife of the abovesaid John Wood, who departed this life October the 9th, 1757, in the 84th year of her age.

Also, THOMAS WOOD, son of John and Arabella Wood, who departed this life November the 8th, 1757, in the 47th year of his age.

On a Portland stone ledger—

Here lieth interred the body of ARABELLA WOOD, daughter of John and Arabella Wood, of Waddon, who died February the 8th, 1738, aged 35 years.

And also the body of JOHN WOOD, son of the abovesaid John and Arabella Wood, who departed this life the 9th of April, 1736, aged 32 years.

Near the entrance to the middle aisle, to the south, on a brass plate, beneath the indents of a cross, between a kneeling figure and a shield*—

* The arms on this tomb, recovered from a MS. in the Lansdowne Collection, were *gu.* two wings conjoined in lure, the tips downwards, *or.* Over all, a label of three points.

Hic jacet Egidius Seymor, qui obiit xxij die
Decembr. a. dni mcccclxxxx cui' aie. ppicict. ds.

To the south of the last, on a rough ledger—

In memory of RICHARD POORE, Esquire, late of the Island of Jamaica in the West Indies, who died the 21st of August, 1788, aged 52 years.

Tho' I my God have oft offended,
May I by Christ be recommended
To thy great mercy and thy love,
To live with thee in heaven above.

At the foot of the last, on a black marble ledger, with arms, a chev. between three eagles' heads erased, in an escutcheon of pretence, three tufts of grass, empaling the same—

Here lieth the body of Mr. JAMES PETTIT, late of Combe, in this parish, Gent., who departed this life the 7th of March, 1724, aged 64.

In the north corner of the chancel, on a black marble ledger

In memory of MARGARET LEE, who died 18th July, 1787, aged 2 years and 7 months.

Also, MARY LEE died August 18th, 1793, aged 8 years and 10 months.

St. Mary's Chancel.

On the east wall is a beautiful monument of white marble, sculptured by the late John Flaxman, R.A., representing an angel bearing up a female. Above the figures are these words:—

Thus
shall the good be received
into life everlasting.

Under—

Sacred to the Memory of
ANN,
The beloved wife of JAMES BOWLING,
of the Borough of Southwark,
(and Daughter of the late Mr. James Harris, of this place),
who, after five days illness only,
exchanged this life for a better on the 26 April, 1808,
in the 25 year of her age.

Bright excellence, with every virtue fraught,
Such may we be by thy example taught;
Pure in the eye of heaven like thee appear,
Should we this hour death's awful summons hear;
Like thee, all other confidence disown,
And, looking to the cross of Christ alone,
In meekness tread the paths thy steps have trod,
And find, with thee, acceptance from our God.

Her husband, under the strongest bonds of affection, has caused this monument to be erected in testimony of his everlasting regard and gratitude to a most affectionate wife, and kind friend.

On a white marble tablet, to the north of the last, is this inscription:—

In the family vault, near this place, are deposited the remains of SAMUEL MARSH, Esq., of Bellemont House, near Uxbridge in the county of Middlesex, who died March 18, 1795, aged 78 years. His affectionate widow has caused this monument to be erected to his memory.

On a like tablet, by the side of the last—

To the memory of Captain John Marsh, of the 66th regiment, who died Feb. 27, 1798, aged 21 years. Also to the memory of Frances Elizabeth Marsh, widow of the late Samuel Marsh, Esq., of Bellemont House, who died October 27, 1811, aged 64 years.

In the north-east corner, on a raised tomb,

formerly railed in, are the indents of a figure and two shields. On a brass plate under the figure is the following inscription:—

Orate pro anima Elys Davy, nuper Civis & Merceri
London, qui obiit iiij die mens' Decembris, Anno Dni
Mill'imo cccclv. cujus anime propicietur Deus. Amen.

Above this tomb, on the north wall, is the following inscription on a brass plate:—

Heare lyes ye body of y^t precious servant
of God, Mr. SAMUEL OTES, Master of Arts &
Minister of the worde in Croydon, whose
Piety, Zeale, & Selfdenyal, are the best Mo-
-nument of his Worth : whose blessed memery
lives & need not words to preserve it. he was
placed there A°. 1643, & deceased A°. 1645, aged 30
yeares, Having lived long, though he dyed young.
R (admire & learne) B.

To the west of the last is a large white tomb, ascended by three steps, bearing the figures of a man in armour, in alto relievo, kneeling before a desk, attended by his five sons, and a woman in the same manner, attended by eight daughters. Over the heads of the women are these initials:—

K. A. M. S. E. A. M. E. M.

Between the figures—

Anno Domini 1568.

Over the heads of the men—

H. W. T. I. P. N.

At the bottom of the tomb is this inscription:—

Tumulus Nicholai Herone, Equitis, sepulti primo die Septem.

The arms on this tomb are—centre shield, quarterly—1, a chev. charged with three cinquefoils between three herons (for Heron); 2, two bendlets; 3, a fesse between three boars' heads, couped; 4, a chev. engrailed between three bugle horns stringed. Empaling 1 and 4, semmè of fleur de lis, a lion ramp., charged on the breast with an annulet; 2, a chev. between three stags' heads cabossed; 3, three martlets. Dexter shield—quarterly: 1 and 4, semmè of fleur de lis, a lion ramp., as above; 2, a chev. between three stags' heads cabossed; 3, three martlets. Sinister shield—quarterly: 1 and 4, Heron; 2, two bendlets; 3, a fesse between three boars' heads couped.

On the Ground.

Near the entrance of this chancel, from the middle chancel, is a rough ledger, with this inscription:—

Elizabeth Butler, the wife
of Francis Butler, Esquire,
was buryed Novemb. 26,
1626.
The said Francis Butler,
Esquire, was buryed the
4th of June, 1648.

Adjoining the last, on the north, is a black marble ledger, with these arms—*sa.* on a fesse *or*, between three bugle horns stringed *arg.*, a de-mi-lion naisant *gu.* between two pheons *az.*, empaling *gu.* a chev. *or*, between three lions ramp. *arg.*—and the following inscription:—

Here lye the body of THOMAS JOHNSON, senr. (late of this parish, and vintner of London), obiit 16 Feb. 1726, æt. 56.

Also JOHN JOHNSON, his brother, ob[t] 14 April, 1721, æt. 52.

And of JOHN, son of the said Thomas Johnson, obiit 24 July, 1723, æt. 16.

By the side of the last, on a black marble ledger—

M. S.
Subtus
In spe Beatæ Resurrectionis
Requiescunt Exuviæ
ANNÆ EYRES Viduæ,
Relictæ Thomæ Eyres, M.D.
Quæ
Obijt, 2° die Martij, A° D[i] 1717°,
Ætatis suæ 78.
Matrona Bona, Justa et Pia.
In cujus perpetuam ac gratam memoriam
Hoc Marmor
Johannes Eyres Filius natu maximus
Mœrens Posuit.

Adjoining the last are the indents of the figures of a man and woman.

At the foot of Johnson, on a black marble ledger—

ELIZABETH, the wife of George ELCOCK, Citizen and Draper of London, departed this life the first of July, 1648.

Likewise the body of GEORGE ELCOCK, husband to the said Elizabeth, who departed this life the 8th of August, 1657.

Adjoining the last, to the north, within the rails inclosing the Font, on a Portland stone ledger—

Here lyeth the body of Mr. CHARLES WESTGARTH, of Unthank, in the county of Durham, who departed this life the first of July, 1733, aged 35 years.

By the side of the last, within the rails, is a white ledger, with arms *az.*, a cross of lozenges *or*, on a chief *gu.* a leopard passant gardant of the second spotted *sa.*, holding in his dexter paw a fleur de lis of the second, empaling a fesse coticed between three cats. The inscription, which is concealed by the font, commemorates—

BENJAMIN DELAUND, who died June 19, 1753, aged 79; MARGARET DELAUND, who died January 2, 1714, aged 78; and RICHARD DELAUND, who died . . . . . .

By the side of the last, and also within the rails, on a black marble ledger—

Here lyeth the body of Mrs. ANN CALLANT, widow, eldest daughter of Thomas Morton, Esq. (of Whitehouse), who departed this life the 11th of February, 1735, in the 72nd year of her age.

Here lyeth the body of JANE CALLANT, the wife of Robert Callant; ob[t] the 29th of October, 1736, aged 52 years.

Also MARTHA, second wife of the s[d] Robert Callant; ob[t] Sept. y[e] 28th, 1741, ætatis 45.

Also the body of ROBERT CALLANT, who died y[e] 7th February, 1764, aged 72 years.

Adjoining the last, by the wall, on a white marble ledger, with arms *or* a lion ramp. regardant *sa.*, empaling quarterly *gu.* and *erm.* in dexter chief and sinister base, a goat's head erased *arg.* attired *or.*

Here lyeth
the body of Mrs. Elizabeth Price,
Wife of Herbert Price, of ye County
of Hereford, Gent., and Daughter to
Thomas Morton, of White-Horse,
in this Parish, Esqr., who departed this
life the 15th day of February, in y 35
yeare of her age, 170½.
Also near this place lyeth three of
their Children (viz.) Jane, Susanna,
and Thomas Price.

Charo viro et natis vixit; Charissimo Christo
Vivat, et æterne huic pax sit et alta quies.

At the head of the last is the indent of a figure, with labels issuing from its mouth.

Adjoining the last, to the west, on a Portland stone—

Also of Mrs. Frances Hutchinson, died 19th July, 1825, aged 84 years.

Adjoining the last, to the north, on a like stone—

C. B. Apthorp, aged iii months xiii days, died ix October, mdcclxvi.

Miss Catherine Hutchinson, died January 22, mdcclxxvii, in the xxivth year of her age.

Mrs. Elizabeth Apthorp, born March 2, mdccxli, died January 28, mdcclxxxii.

Mrs. ELIZABETH HUTCHINSON, eldest daughter of Lieut. Gen. Will. Shirley, died March 22, MDCCXC, aged LX.

WILLIAM HUTCHINSON, Esq., born at Boston, in America, died 8th February, 1797, aged 54 years. He was a man of strict probity and true honour, and a zealous, faithful friend, an affable and kind relation, and a worthy member of the Established Church. He entertained the highest and most uniform principles of loyalty, to which he sacrificed his private interest. As agent to the Island of Antigua, and in other public stations, his conduct received the fullest approbation. He merited and enjoyed universal esteem.

Before the tomb of Sir Nicholas Heron, on a black marble ledger, is this inscription:—

Here lyes ye body of
MORREN HARBIN, Citizen &
Dyer of London, who
departed this Life ye 22th of
October, ano. Dom. 1680,
aged 55 yeares:
Fourteen days before whose
death, Hellen his wife departed
this life, aged 44 years,
and lieth both here interred.

Adjoining the last, on a like stone—

Here lyeth the body of
EDMOND the Sone of MORREN HARBIN,
and Hellen his wife, who departed
this life the 8th day of July,
Anno Domini, 1682,
in the 19th year of his age.

St. Nicholas' Chauntry.

On the east wall is a monument, bearing, under a recessed arch, the effigies of a man in a black gown, kneeling before a desk; and these arms—*arg.* three crosses patée, ends fleury *gu.* each charged with five bezants; in a canton of the second, a conger's head couped in pale *or.* —Crest, a lion's jamb erect and erased *or*, holding a like cross and bezant fitchey. Motto, "Meliora manent." Over his head is this inscription—

Ossa Michaelis sunt hic sita
Murgatroidi. Da, pia posteritas,
vere quiete cubent.

Beneath his feet, on a black marble tablet, is this inscription—

Michael Murgatroid Eboracensis, Richardi Gascoigni armigeri alumnus, olim Collegii Jesu apud Cantabrigienses socius, postea Johanni Whitgift Archiepiscopo Cantuariensi ab epistolis, inde ejus familiæ Censor sive Contrarotulator, denique Dispensator sive Senescallus, et ad Facultates in alma Curia Cantuariensi Commissiarius: vixit annis 56, mensibus 4, diebus 12; obiit tertio die Aprilis, anno salutis humanæ 1608.

On a monument in the south-east corner, greatly resembling that of Archbishop Grindall's, is the recumbent effigies of a churchman in sable robes, with his hands in the act of prayer. The arms on this tomb are—Centre shield, the arms of the see of Canterbury, empaling *arg.* on a cross fleury *gu.* five bezants: dexter shield, the arms of the see of Worcester; sinister shield, the arms of the deanery of Lincoln, both empaling the same. On the sarcophagus are the arms of the see of Lincoln, the colleges of Trinity, Pembroke, and Peter-house.

At the top of the monument is the following inscription*:—

Post tenebras spero lucem.

Above the figure—

WHITGIFTA Eborum Grimsbeia ad littora nomen
Whitgifta emisit. Fœlix hoc nomine Grimsbei.
Hinc natus, non natus ad hanc mox mittitur hospes
Londinum. inde novam te, Cantabrigia, matrem
Insequitur, supraq. fidem suavi ubere crescit.
Petro fit socius, Pembro, Triadiq. magister,
Fitq. Pater matri, Cathedræq. Professor utriq.
E Cathedra Lincolna suum petit esse Decanum,
Mox Wigorn petit esse suum, fit Episcopus illic:
Propræses Patriæ, quo nunquam acceptior alter.
Post annos plus sex summum petit Anglia patrem;
Plusquam bis denos fuit Archiepiscopus annos.
Charior Elizæ dubium est, an Regi Jacobo:
Consul utriq. fuit. Sis tu Croidonia testis
Pauperibus quam charus erat, queis nobile struxit
Hospitium, puerisq. scholam, dotemq. reliquit.
Cœlibis hæc vitæ soboles quæ nata per annos
Septuaginta duos nullo enumerabitur ævo.

* The inscriptions on this tomb were written by Dr. Benjamin Charior, one of his Grace's chaplains.

Invidia hæc cernens moritur, Patientia vincens
Ad summum evecto æternum dat lumen honori.

A little lower, the two following verses, in juxtaposition—

Magna Senatoris sunt nomina, pondera & æqua
Nominibus, quem non utraq. juncta premunt?
Præsulis accedat si summi nomen ad ista
Pondera quis ferat, aut perferat illa diu?

Pax vivo grata est, mens recti conscia pacem
Fert animo, hæc mortem non metuisse dedit.
Mors requiem membris, animæ cœlestia donant
Gaudia; sic potuit vincere qui patitur.

Beneath the figure—

Gratia non miro, si fit divina Johannis
Qui jacet hic solus credito gratus erat.
Nec magis immerito Whitgiftus dicitur idem;
Candor in eloquio, pectore candor erat.
Candida pauperibus posuit loca candida Musis:
E terris moriens candida dona tulit.

Adjoining the last, on the south wall, is a tomb, which is presumed to commemorate Thomas Warham, Esq., who died at Haling, 1478, and who ordered his body to be buried in the chapel of St. Nicholas, before the image of our Lady of Pite. The tomb, which is inserted in the wall, is divided at its base into three quatrefoil panels, each containing a shield of arms—viz. centre shield, quarterly, 1 and 4, *gu.* a fesse *or*, in chief a goat's head couped *arg.*, attired of the second; in base three escollops of the third, with a mullet *sa.* for difference, all within a bordure of the third, for Warham; 2 and 3, two bars: dexter shield, Warham: sinister shield, two bars empaling the same. Over the tomb is an abstrusely pointed arch, surmounted by a richly sculptured cornice; above which, on the wall, is a shield with mantling and helmet, bearing the arms of Warham, and two bars quarterly; but the crest (an armed arm holding a sword) has disappeared. Crowning the angular pillars flanking the arch, are two shields of arms, viz. 1, Warham; 2, two bars, as before. The soffit of the arch is divided into trefoil headed panels, and in its recess are the indents of the figures of a man and woman kneeling, with labels issuing from their mouths; which, with every other inscription, have been sacrilegiously torn away, probably during the Rebellion. On this tomb are two helmets.

Adjoining the last is the well-known monument of Archbishop Sheldon, representing the recumbent effigies of the prelate in his archiepiscopal robes and mitre. His left hand sustains his head, and in his right is a crosier. The figure is of statuary marble, and is beautifully sculptured. In the panels of the black marble altar tomb on which the archbishop reposes, is some finely carved osteology. Above the figure is the following inscription, surmounted by two cherubs supporting a shield of arms—viz. *arg.*

on a chev. *gu.*, three sheldrakes of the first on a canton of the second, a rose of the last, empaled by the arms of the see of Canterbury. Motto, "Fortiter et Suaviter*".

Hic jacet
GILBERTUS SHELDON
antiquâ SHELDONIORUM familiâ
In agro STAFFORDIENSI natus,
OXONII
bonis literis enutritus,
S. Stæ Theologiæ Doctor insignis;
Coll. Omnium Animarum CUSTOS prudens et fidelis,
Academiæ CANCELLARIUS MUNIFICENTISSIMUS,
Regii Oratorii Clericus,
CAR. I^MO B^MO MARTYRI CHARISSIMUS;
sub Serenissimo R. CAROLO II^DO,
MDCLX, magno illo INSTAURATIONIS anno,
Sacelli Palatini Decanus,
LONDINENSIS EPISCOPUS;
MDCLXII, in Secretioris CONCILII ordinem
cooptatus;
MDCLXIII, ad dignitatis ARCHIEPISCOPALIS apicem
evectus.
Vir
Omnibus Negotiis PAR, omnibus Titulis SUPERIOR,
In Rebus adversis MAGNUS, in prosperis BONUS,
Utriusque FORTUNÆ Dominus;
PAUPERUM PARENS,
LITERATORUM PATRONUS,
ECCLESIÆ STATOR.

* This splendid monument, so universally admired, which Evelyn terms "of a stately ordnance," and estimates its cost at from 700*l.* to 800*l.*, was the work of Joseph Latham, the city mason, and was entirely finished by English workmen, circa 1683.—Vide the Present State of England, 1683, p. 152.

De tanto Viro
PAUCA dicere non EXPEDIT, MULTA non OPUS EST;
Norunt PRÆSENTES, POSTERI vix credent:
OCTOGENARIUS
Animam PIAM et Cœlo MATURAM
DEO REDDIDIT
v Id. IX Bris
MDCLXVII.

On a neat white marble tablet, affixed to the wall, nearly opposite the last—

Beneath are deposited the remains of the most reverend JOHN POTTER, D.D., Archbishop of CANTERBURY, who died October x, MDCCXLII, in the LXXIV year of his age.

Directly above the last, on a white marble tablet, with arms—Per pale, *erm.* and *or,* a man in armour unhelmeted proper, his dexter hand resting on a rock. In a canton *vert,* a ship of the second—

Extra hæc mœnia sepultum
Quod mortale fuit
GEORGII LEONARDI STEINMAN,
Armigeri,
Filii unigeniti
Leonardi Steinman,
De Sti Gallen oppido, Helveticis civitatibus;
Natus in oppido supra memorato,
I Martii, MDCCLVIII.
Mortuus in hac parochia,
IV Januarii, MDCCCXXX.

Ibidemque sepulta
LOUISA BASTIN,
filia natu minima
Georgii Leonardi & Susannæ Steinman,
Conjux Philippi Henrici Byrne, Armigeri.
Decessit XIII Julii, MDCCCXXVIII.
ætat. XXXIII.

On the Ground.

Between the pews adjoining the east wall, on a black marble ledger—

Here lyeth the body of
The most reverend DR. THOMAS HERRING,
Archbishop of CANTERBURY,
who died March 13, 1757, aged 64.

Adjoining the last, to the north, on a like ledger, is the following inscription, now partly hid by the pews—

Underneath lyeth interred (near the remains of her parents) the body of Mrs. DOROTHY PENNYMAN, relict of Sir James Pennyman, of Thornton, in the county of York, Baronet, and one of the daughters and co-heirs of Dr. William Wake, late Lord Archbishop of Canterbury. She died the 2nd day of December, 1754, aged 55 years.

At the head of the last, on a like ledger—

PETER CHAMPION, Esq., died May 27, 1758, aged 75.

Adjoining the last, to the north, on a like ledger, with arms—*erm.*, three trefoils *sa.*, empaling *az.* on a fesse between three martlets *arg.* three cross croslets of the first—

In memory of CATHERINE CHAMPION, wife of Peter Champion, who died after a tedious illness, suffered with resignation, November the 14th, 1750, in the 63rd year of her age; having always deserved well of her husband and children.

Adjoining the last, on a like ledger, with arms, *gu.* a fesse vairé between three ravens *or*—

In memory of ANTHONY WALLINGER, of London; he died June the 4th, 1728, in the 90th year of his age.

Adjoining the last, on a like ledger, with arms *az.* on a fesse between three martlets *arg.* three cross croslets of the first—

Here lieth the body of Mrs. Elizabeth Wright, who died March 15th, 1784, æt. 59 years.

Before the tomb of Archbishop Whitgift, on a Portland stone—

Here lye ye bodies of 2 sons & 1 daughter of Henry & Hannah Mills, who died in their infancy:

Mary, buried June 26, 1716.
John, —— Nov. 17, 1717.
Hannah freeman, June 6, 1724.

Before the tomb of Warham, on a marble ledger—

Here lieth the body
of
Sir Joseph Sheldon, Kt, some time
Ld. Mayor of London, the eldest
son of Ralph Sheldon, Esqr,
who was
the elder brother of Gilbert Sheldon, Ld Archbishop
of Canterbury.
He left issue two daughters, Elizabeth & Ann,
and
died Augst ye 16°, 1681,
in the 51st year of his age.

Adjoining the last, on a like ledger—

Here lieth the body of Daniel Sheldon, Esq., son of Ralph Sheldon, Esq., elder brotr of Gilbert Sheldon, Ld. Archbishop of Canterbury. He died 14 Feb. 1698, aged 65, leaving issue one son, Gilbert, and two daughtrs, Judith and Mary.

Adjoining the last, on a veined marble ledger—

Here lyes interr'd the Body of JUDITH SHELDON, who died Decr. 6th, Anno Domni 1725, Aged 47 years. She was daughr of Daniel Sheldon, Esq., who also lyes interr'd near this place.

At the head of Daniel Sheldon, on a marble ledger—

Here lieth the body of ROGER SHELDON, Esq., son of Ralph Sheldon, Esq., who was the elder brother of Gilbert Sheldon, Ld Archbishop of Canterbury; he died unmarried, 30th May, 1710, aged 71.

At the head of Judith Sheldon, on a like ledger—

JOHN DURAND, Esq., died at Woodcot Lodge, near Carshalton, Surry, XIX July, MDCCLXXXVIII, in the LXX year of his age.

At the foot of Durand, on a Portland stone ledger—

Here lyeth the body of JOHN MATTOCK, citizen and haberdasher of London: obiit June 9, 1720.

Also Mr. RICHARD MATTOCK, apothecary, who died February the 25th, 1720-21, aged 26.

A little to the west of Durand, on a large black marble ledger—

Sub hoc marmore deposita sunt corpora
RACHELIS uxoris THOMÆ BRIGSTOCK, Armig.
quæ obiit XVII Kal. Aug. A.D. 1756; ætatis ejus XLVI.
Thomæ Brigstock supra nominati,
Caroli filii Thomæ Brigstock junioris et Annæ Papwell
conjugis ejus, nepotis Thomæ et Rachelis.
Quorum
Thomas obiit X Kal. Mart. 1771, ætatis LXIV.
Carolus quatuor hebdomadum infans X Kal. ejusdem
Mensis et anni.
Avus et Nepos eodem die sepulti sunt.

Hic etiam jacet
Ricardus Papwell Brigstock,
Caroli frater infans,
Qui XII hebdomadas natus decessit
VII Idus Decembris A. 1785.
Necnon
Anna Rachel Brigstock,
filia Thomæ Brigstock et Annæ Papwell uxoris,
quæ obiit VIII Nonas Maii 1787, ætatis A. XIII.
Et
Thomas Brigstock armiger,
qui obiit
Pridie Idus Octobris, A.D. 1787, ætatis XLIX.
Esto Fidelis usque ad mortem,
Et Dato tibi coronam Vitæ.
Thomas Brigstock, filius Thomæ Brigstock
et Annæ Papwell, uxoris, qui obiit
XXVII Octob. 1792, ætatis XVI.

At the foot of the pillar, at the entrance of the south aisle, on a black marble ledger, with arms —*sa.* three horse buckles *arg.*, with an annulet for difference. The inscription, which is partly hid, is as follows:—

Here lieth interred the body of HENRY MARTIN, late Citizen and Grocer of London, son of Thomas Martin, of Bowsham in Oxfordshire, gentleman, and Lucie his wife, who departed this life February the 27th, 1603, aged forty-two years and two months.

Nave.

On an elegant marble column, supporting an urn, on the south wall of the east end of the nave, designed by Glover, the author of "Leonidas," is the following inscription—arms *arg.* on a chev. *sa.*, three mullets *or*, between three pellets *gu.* for Bourdieu, empaling per pale *az.* and *gu.*, a cross engr. of the first—

Near the remains
of his beloved wife PHILIPPA BOURDIEU,
this monument was erected
by James Bourdieu of Comb, in the county of Surrey, Esq.,
whom, with ten children,
the objects of her long and unwearied care,
she left behind her,
under the most unfeigned affliction
at their common irreparable loss;
she died
at the age of 50, on the 24th of June, 1780.

On a marble tablet, under the above—

Near this place
are deposited the remains of
JAMES BOURDIEU, ESQ.,
of Coombe, in the county of Surrey, who died on the 3rd of
November, 1802, in the 90th year of his age.

In the same situation as Philippa Bourdieu's monument, on the opposite side, is a column of white marble, supporting a funereal urn, with the following inscription—arms Bourdieu, empaling *gu.* a fesse vairé; in chief an unicorn passant, between two mullets *or;* all within a bordure of the last, pellettée—

Sacred
To the Memory of
MRS. ANNE BOURDIEU,
Wife of John Bourdieu, Esq.,
of Golden Square, London.
She departed this life
the XXIII of March
MDCCXCVIII
aged XXXI years.

A virtuous daughter and a sister kind,
A tender mother, and a wife refin'd,
Who all the various dues of life sustain'd,
Inspir'd by wisdom and in honour train'd,
Lies here entomb'd, here virtue, beauty, grace,
Ready for heav'n, have run their earthly race;
Yet, to the shorten'd course of youth confin'd,
She shew'd but glimpses of her glorious mind;
Where multitudes of virtues pass'd along,
Each moving onward in the lovely throng,
To kindle admiration, and make room
For greater multitudes that were to come;
But her vast mind, rich with such gifts divine,
In heaven's eternal year alone could shine.

On the Ground.

On a brass plate inlaid in a Portland stone ledger, is the following inscription:—

Here under are conteined the bodyes of
Thomas Parkinson, late farmer of ye Parsonage
of Croydon, and Elizabeth his wife, which
Thomas deceased ye 7th day of September, 1603,
And Elizabeth the 30th of January, 1594.

On a large black marble ledger, to the west of the last, is the effigies, in brass, of a man, and the indents of the figures of a woman and child. The inscription plate is gone, but the following are the words it contained:—

Here lieth interred the body of John Packington, late
of the Parsonage of this Towne of Croydon, who deceased the xxii day of June An. Dom. 1607, leving issue one
Onely childe Henry Packington by Anne his wife, who,
yet surviving, at her decease appoynteth heare
her place of buriall*.

* Vide Aubrey's Antiquities of Surrey, Vol. 2.

On the same stone, a little lower—

Curteous Reader, knowe that here doth ly
A rare example of true pietie,
Whose glorie 'twas to prove herselfe in life
A vertuous wooman, and a loyall wife.
Her name to you obscurely Ile impart
In this her Anagrame no arme but Hart;
And least you joyne amis & soe loose yᵉ name
Looke underneathe & you shall find yᵉ same.
Martha Burton yᵉ wife of Barnard Burton Esqr.
deceased yᵉ 20h day of November, & was buryed
yᵉ 26h day, Anᵒ. Dni. 1668.

To the west of the last, on a Portland stone ledger—

Here lieth Interred the Body of M. Ann Catherine Berne, who departed this life April the 10, 1758, aged 5 years.

Adjoining the above, to the south, on a like stone—

In memory of Mrs. Anna Maria Berne, who died April
in her 60th year.

Adjoining the last, to the west, is a like stone, covering the remains of James Bourdieu, Esq., before mentioned.

Adjoining the above, to the south, is a like stone, covering the remains of Phillippa, his wife, also before mentioned.

Adjoining the last, to the south, on a like stone—

Here lyeth the bodies of Andrew Smith, who departed this life 23rd of June, 1755, aged 48 years.

Also Hannah his wife, who departed this life the 12th of December, 1735, aged 73 years.

To the west of the last, on a black marble ledger, is the brass effigies of a woman; and, on a brass plate beneath, these words—

Elizabeth daughter of John Kynge and Clemence his wyfe the wyfe of Samuel Ffynche, unto whom she bare three sonnes & two daughters, and deceasynge the xiij daye of November, here lyeth interred; anno Dni 1589, ætatis suæ 21.

To the west of the last, in the middle of the aisle, on a Portland stone ledger—

Here lyeth interred the body of Miss Abigail Cooke, daughter of Mr. Philip Cooke and Mary his wife, who departed this life the 16th day of December, 1776, aged 30 years.

Also Mrs. Mary Cooke, mother of the above-named Abigail Cooke, who departed this life the 10th of January, 1769, aged 60 years.

Adjoining the last, to the west, on a like stone—

"That if thou shalt confess with thy mouth the Lord Jesus, and shalt believe in thine heart that God hath raised him from the dead, thou shalt be saved."—Romans x. ch. ver. 6.

Underneath this stone are deposited the remains of Mrs. Elizabeth Hetherington, wife of Mr. Theophilus Hetherington, who died the 28th day of April, 1768, aged 75 years.

Adjoining the last, to the west, on a black marble ledger—

Stephen Galhie, Gent.
died September 16, 1772, in the 70th year of his age.

Adjoining the last, to the west, on a Portland stone ledger—

In memory of George Reavely, Esq., who departed this life January 17, 1780, aged 65 years.

Adjoining the last, to the north, on a like stone—

In memory of Mrs. Rose Belgrave, departed this life the 2nd of September, 1780, aged 86 years.

North Aisle.

On a neat white marble tablet, on the wall near the pulpit, is the following inscription:—

Sacred
To the memory of
Henry Richard Raven, Esq[re],
of this parish,
who died September 27th, 1831, aged 49 years.
His remains are deposited
in the church of St. Vedast.
This tablet is erected by his widow
as a tribute of affectionate respect to the
memory of a husband whose loss she can
never cease to lament.

On a copper plate, affixed to the north wall, with arms—*or*, a buck trippant *gu.* in a canton of the last, a ship *arg.* On an escutcheon of pretence, quarterly *arg.* and *sa.* on a bend *gu.*, three lions passant guardant *or*, empaling the same.

Mrs. Mary Ann Parker died September ye xviii[th], mdccxxxii. Aged lxx years.

On the Ground.

At the entrance of the aisle, from the west, on a black marble ledger—

Memoriæ sacrum:
To the pious Memorye of his religious Father,
Ralph Smith, who deceased the 26
of Sept. 1639, aged 83, Thomas
Smith did lay this marble,
as a grateful testi-
monye of his
Filial Duty.

So well thou lov'st God's House, tho', beinge blinde,
Thou came oft thither, lighted by thy mind;
Where thou didst offer such a sacrifice
As few do now present that have their eyes;
A bleeding harte of sinne in sorowe Dround,
Sustain'd by Hope and with Devotion cround;
Therefore thou dost deserve an abler Pen,
Whose spritely Lines mighte stir up zeale in men;
To write thine Epitaph, I am sure of this,
What thou dost want in Words thou hast in Blisse.

To the south of the last, on a Portland stone ledger—

Susanna Copley died October 25th, 1785, aged 9 years.
Jonathan Copley died Nov. 9, 1785, aged 3 years and 8 months.

Adjoining Smith, on the west, on a black marble ledger—

Here lieth the body of
Marmaduke Wyvel, Esqr;
and one of ye King's Majie^{ts}
Pentioners, second Sonne
to Sr Marmaduke Wyvell, of
Cunstable-Burton, in York-
shiere, Knight and Barronet,
who dyed ye xxth of August
1623, aged 58.

On the same stone, lower, in small Roman letters, is this inscription:—

Juxta hic jacet
In spem certam resurgendy* Depositum
Corpus Marmaduci Wyvell, Armigeri,
Filii secundo-geniti Dni Marmaduci
Wyvell, de Cunstable-Burton, in Agro
Eboracensi, Equitis & Baronetti.
Ibidemque reconduntur Corpora Marmaduci et Judithæ filiæ ejusdem
Marmaduci Wyvell, supra nominati:
Beati sunt pulveres,
Quibus promittitur a Christo
Resurrecto ad gloriam
in Rego suo:
Adveniat cito ora tu etiam Lector,
Obiit 2 die Januarii 1678, ætat. suæ 69.

Adjoining the last, on the west, on a black marble ledger—

Here lyeth buried
Elizabeth the daughter of Mr. Roberte Crowe & Catherine his wife, which Elizabeth deceased in the year of our Lord 1638.

Adjoining the last, on the south, on a white marble lozenge, inserted in a Portland stone ledger—

M. S.
in
Spem certam
Resurrectionis,
Hic jacet
Cornelius Clifton,
Juvenis eximiæ spei.
Obiit 15 Maii,
Anno { Domini MDCIX,
Ætatis suæ
xx^mo }

* Sic. orig.

Adjoining the last, on the west, on a like stone—

In memory of PETER HARRISON, Esq., of this parish, who departed this life November 22, 1755, aged 70 years.

Adjoining the last, on the west, on a like stone—

Mrs. SARAH GIBSON, who died June the 1st, 1761, aged 72 years. Also WILLIAM GIBSON, husband of the above Sarah, died the 19th of April, 1773, aged 87 years.

At the head of Crowe are the brass figures of a man and woman, but the inscription plate is gone; some distance from which is another stone, with the indents of the figures of a man and woman, and the indents of two shields of arms. That over the woman contained formerly a lion rampant.

Westward of the last, on a veined marble ledger, with arms, (now concealed by a fireplace), quarterly, 1 and 4 *or*, a buck trippant *gu.*, on a canton of the second a ship *arg.*; 2 and 3 *arg.*, a mullet *sa.*—

Here lyeth the body of CHRISTOPHER, son of John and Bathsheba PARKER, who died the 7th of October, 1711, aged 3 weeks.

Also HENRY the son of the aforesaid, who died February the 2nd, 1717, aged 11 months and 2 days.

Likewise the body of the aforesaid JOHN PARKER, who died the 16th of June, 1740, aged 52.

Also the body of BATHSHEBA PARKER, wife of the above John Parker; she died May 6, 1763, in the 84th year of her age.

South of the last, on a Portland stone ledger—

Here lieth interred two daughters and a son of FREDERICK BURR,

Esq., by Catherine his wife, viz. Catherine, who died January 1, 1734-5, aged 8 weeks: Also Petronella, who died April 18, 1739, aged one year, seven months, and seven days: Also Samuel Alexander, who died May 1, 1732, aged seven months and sixteen days.

Westward of the last, on a black marble ledger—

Here lieth the body of Mrs. ELIZABETH WHITAKER, late wife of Edward Whitaker, Esq., who died September 1, 1727, aged 25 years.

And also ELIZABETH their daughter, who died April 18, 1728, aged 11 months.

Adjoining the last, on the west, on a like stone—

In memory of JOHN ELDERTON, late of Lincoln's Inn, and of this parish, gentleman, who died the 5th of August, 1782, in the 53rd year of his age, beloved and regretted by all his friends.

Adjoining the last, on the west, on a Portland stone ledger—

Here lyeth the body of Mr. JOHN BAYNHAM, who departed this life the 20th day of January, 1779.

South Aisle.

On an oval white marble tablet affixed to the wall, opposite Archbishop Sheldon's monument, is the following inscription:—

Beneath this place were deposited the remains of THOMAS BRIGSTOCK, Esq.; he died of a decline, 27th October, 1792, in the 17th year of his age. If a suavity of manners and goodness of mind could have preserved his life, he had not now been numbered among the dead.

Near the last, on the same wall, is a handsome white marble tablet, bearing the representation of a rose and pruning knife, with the words—"*Flos nimium brevis*," and the following inscription:—

To the revered memory
of
Frances, fifth daughter
of the late Samuel Davis,
of Birdhurst Lodge, in this parish, Esq,
Born 7th June, 1810, deceased 10th May, 1828,
This tablet,
in the absence of her own surviving brothers,
is erected
by the husband of her eldest sister,
who, having known her from childhood,
offers this last tribute
of brotherly love.

On a brass plate, on the south wall—

Here under lieth Buried the bodie of Franc.
Tirrell, somtimes Citizen and Grocer of London.
He was a good Benefactor to the poore of
divers Hospitalls, Prisons, and Pishes of London,
and to the continuall releif of the poore
Freemen of the Grocers. He gave to this Pishe
200*l.* to build a newe Market-house, and 40*l.*
t° beutifie this church, and to make a new
Saintes Bell*. He died in September 1600.

On a handsome white marble tablet, on the same wall—

In memory of THOMAS BAINBRIDGE, of Croydon Lodge, Esquire, who departed this life, January 8th, 1830, in the 81st year of his age.

* One John de Aldermaston, who was buried in this church in 1403, left, by will, twenty sheep, for the purchase of a new saints' bell. Vide Reg. Arundel, fol. 212 b.

On the Ground.

On a black marble ledger, with arms—*vert*, a bend *arg.* cotised *or*, with a crescent for difference; empaling . . . . .

Under the pew lieth interred the body of Mrs. ANN SOPHIA PEERS, late wife of Mr. Richard Peers, who departed this life, April the 10th, 1766, aged 42 years.

Also the above RICHARD PEERS, Esq., Alderman of London, who departed this life June the 25th, 1772, aged 72 years.

Also THOMAS PEERS, Esq., brother of the above, who departed this life November 15, 1765, aged 55 years.

To the south-east of the last are the indents of the figures of a man and woman.

To the west of the last, on a Portland stone ledger—

Underneath this stone are deposited ye Remains of Mrs. CHRIST. FENWICK, wife of Edward Fenwick of South Carolina, Esq., and last surviving Daughter of the late Colonel John Steurt, died November ye 6, 1785, aged 33 years.

At the end of the aisle, on a like ledger—

H. W. Died XXIInd Janry MDCCLXXXII. WILLIAM WELBANK, Esq., Died XVIth Octob. MDCCXL, in the LIst year of his age.

Adjoining the last, to the north, on a like ledger—

To the memory of JANE, wife of THOMAS BREWSTER, of this parish, who departed this life November 1, 1783, aged 38 years.

To the east of the last, on a black marble ledger—

The Honourable JAMES DUGLASS, Esq., Major General of his Majesty's Forces, lyes buried here; he died April the 10th, 1748, aged 75 years.

On a black marble ledger—

To the memory of Mrs. Ann Elizabeth Wilson, widow, departed this life the 28th of May, 1777, aged 77 years.

Cross Aisle.

On a brass plate on the wall, above the north door, is this inscription:—

Near this place are deposited the remains of Joseph Wilks, Esq., of Measham, in the county of Derby, who died May 24th, 1805, aged 73.

On a handsome white marble tablet, on the west wall—

Sacred to the memory of Robert Chatfield, Esq., who departed this life the xxi of May mdcccxi, aged lxiii years. Also of Ann, relict of the above, who departed this life the xxviii day of September, mdcccxxx, aged lxxv years.

On the Ground.

At the entrance of the aisle, on the south, on a black marble ledger—

Here lyeth interred the body of Mrs. Mary Whitehill, late wife of Mr. Thomas Whitehill, of this parish, who departed this life the 22nd of May, 1781, aged 52 years.

Adjoining the last, on a Portland stone ledger—

James Wilkinson, Esq., Captain of Dragoons, died April 7th, 1769, aged 49.

Adjoining the last, to the south, on a like ledger—

Mary Smith, daughter of the late Edward King, Esq., of Bromley, Kent, and wife of the Revd. John Smith, A.M., Rector of Weybridge, Surry, died XII day of September, MDCCLXXXVIII, in the LVIII year of her age.

Adjoining the last, to the west, on a like ledger—

Here lieth the body of the Rev. JOHN VADE, vicar of this parish, who died the 9th of June, 1765, aged 42 years.

Also, the remains of Miss MARY VADE, his daughter, who died 28th March, 1790.

Likewise, Mrs. ELIZABETH VADE, relict of the aforesaid Rev. John Vade, late vicar of this parish, and of St. Nicholas, Rochester, Kent, who died 23rd July, 1800, aged 80.

To the north of the last, on a like ledger—

In memory of WILLIAM GODFREY, who died August the 3rd, 1770, aged 9 years.

On a Portland stone ledger, at the north end of this aisle—

In memory of MARY CHATFIELD, wife of Allen Chatfield, who departed this life September the 18th, 1761, aged 39 years and 9 months.

Also, the above named ALLEN CHATFIELD, who departed this life the 30th of April, 1772, aged 60 years.

To the north of the last, on a black marble ledger—

Here lyeth interred the body of ROGER DRAKE, Esq., who departed this life June the 20th, 1762, aged 64 years.

Adjoining the last, to the north, on a like ledger—

Here lieth interred the body of ROGER DRAKE, who departed this life January 23, 1770, aged 22 years.

Adjoining the preceding, to the north, on a like ledger—

Here lieth interred the body of BEESTON DRAKE, who departed this life June 14, 1764, aged 21 years.

Belfry.

On a Portland stone ledger—

MRS. SARAH HEATHFIELD died the 17th of February, 1772, aged 61 years.

On a like ledger, adjoining the last, on the north—

Here lyeth the body of JOHN HEATHFIELD, Esq., who departed this life the 8th day of April, 1743, aged 73 years.

Also of ELIZABETH, his widow, who died the 7th of October, 1748, aged 77 years.

Adjoining the last, to the south, on a like ledger—

Here lieth the body of MARY, the late wife of JOHN HEATHFIELD the younger, Gent., who departed this life the 11th day of June, 1741, aged 44 years.

Also, the above named JOHN HEATHFIELD, Esq., died the 14th of November, 1776, aged 78 years.

Adjoining the last, to the south, on a like ledger—

Here lieth the Body of
WILLIAM HEATHFIELD, Esq.
of London, Grocer,
who died Dec. 12, 1791,
aged 56 years.

On the same stone—

Hic . . . . .
Johannis Heathfield, M. A.
nuper Vicarii Eccle . . .
Northaw in Comitatu Hert.

Qui per annos quinquaginta officio
. . . . . . Doctrina orthodoxi
Vitæ inculpabilis Morti resignat . . .
Ob. VII Feb. Ann. Dom. MDCCCV.
Ætat. LXXVII.

Adjoining the last, to the north, on a like ledger—

Here lie the bodies of three daughters of JOHN and MARY HEATHFIELD, viz.—

JANE, died December 19th, 1727, aged three months.
MARGARET, died March 8, 1729, aged one month.
ELIZABETH, died March 27, 1731, aged two months.

Above this row, to the south, on a like ledger—

In memory of ROBERT RIDLEY, son of Thomas and Mary Ann Ridley, departed this life December 11, 1799, in his fourth year.

Also, of JOHN RIDLEY, son of Thomas and Mary Ann Ridley, departed this life December , 1804, aged 16 years and months.

Adjoining the last, to the south, on a like grave-stone—

Sacred to the memory of Mrs. KIMBRA RICHARDS, died January the 21th, 1782, aged 52 years.

Not lost—but gone before.

To the south of the last, on a like ledger—

JOHN CHATFIELD, Esq., of this Parish, departed this life May . . . . aged 56 years.

Sacred to the memory of ANN CHATFIELD, wife of the above John Chatfield, Esq., who departed this life April 4th, 1819, in her 59th year, a sincere Christian and an affectionate wife.

Gallery.

On a large handsome veined marble monument in the north gallery, with arms *or*, a buck trippant *gu.* on a canton of the last, a ship *arg.*, in an escutcheon of pretence *arg.*, a mullet *sa.* empaling the last—

Sacred to the memory of John Parker, Esq., formerly of London, who died the 6th of March, 1706, aged 46 years, and is here interred.

Also of

Elizabeth, his relict, who died the 10th of August, 1730, aged 70 years.

This pair, whilst they lived together, were
A pattern for conjugal behaviour;
He a careful indulgent husband,
She a tender engaging wife;
He active in business,
Punctual to his word,
Kind to his family,
Generous to his friend,
But charitable to all;
Possest of every social virtue.
During her widowhood,
She carefully and virtuously
Educated five children,
Who survived her:
She was an excellent œconomist,
Modest without affectation;
Religious without superstition;
And in every action behaved
With uncommon candour and steadiness.

EPITAPHS FORMERLY IN THE CHURCH,

But now either lost, or concealed by the erection of Pews.

The following memorandum of inscriptions is extracted from a MS. collection of epitaphs, dated 1610, formerly in the possession of Nicholas Charles, Esq., Lancaster Herald*:—

In ye Queere,

MARGARET MORTON, daughter of [William] Woodford, cosyn and heire to Raffe Woodford, of the countie of Leicester, died 1507.

Sr. ROBERT MORTON, Knight, ... and servant to King Henry 8, ob. 1514. His wyfe† was [daughter] and heire to [John] Twinyhoe.

Arms, quarterly, *gu.* and *or*, in the dexter chief and sinister base a goat's head erased *arg.*, attired of the second; on a chief *erm.* three bezants, each charged with an escallop of the first; over all, a label of three points.

THOMAS HERON, Esquire, died 1544; his wyfe was Elizabeth, d. & co-heire to Wm. Bond, clarke of the grene cloth.

This ledger contains the brass effigies of a man in armour, his wife, four sons, and seven daughters. There are five shields of arms, two and

* Lansdowne MSS. No. 874, p. 64.

† I find by the pedigree of this family, that the *wife* of Sir Robert Morton was Jane, eldest daughter of Nicholas Warham, brother of the archbishop of that name, and that his *mother* was Dorothea, daughter and heir of John Twyniho, Esq., of Cirencester.

one above and two beneath the effigies. The first shield contains the arms of Heron. The second (destroyed). The third—Heron; two bendlets, in sinister chief a cross crosslet, pale ways; empaling quarterly, 1 and 4, a fesse betw. three boars' heads couped, 2 and 3, a chev. eng. betw. three bugle horns stringed. The fourth—Heron, empaling quarterly, 1 and 4, two bendlets (as above), 2 and 3 quarterly, 1 and 4, a fesse (as above), 2 and 3, a chev. (as above). The fifth—Heron*.

Middle Chancel and Nave.

Here lyeth WILLYAM HERON, Esquyer and Justys of the Peace, and ALSE his wyfe; which Willyam deceased the iiij daye of January, in the yere of our Lord MCCCCCLXII; whose soule God take to hys mercy. Amen.

This ledger contains the indents of the effigies of a man in armour, and a woman, with the indents of two shields of arms.

On a black marble ledger, with arms—a chev. murally betw. three foxes' heads erased.

Here lyeth interred ye Body of Mrs. SUSANNA LEGATT, ye Wife of Mr. Gorge Legatt, Citizen & Dry-Fishmonger of London, the onely Daughter of Mr. Richard Shallecross, of ye parish of Croydon, Yeoman, aged 24 years, leveing one Son. She departed this Life ye 9th day of September, in ye yeare of our Lord God, 1679.

* Relying upon the general accuracy of the Heralds, in p. 49, we

On a like stone, with arms, a cross saltire—

SARA, the wife of Jonathan Andrews, of London, Merchant, died the 1 of October, 1664.

On a rough marble, with arms, in a lozenge *arg.* a cross saltire *gu.* betw. four eagles displayed *az.*

Here lieth interred the body of the truly pious and singularly accomplish'd Lady Dame RUTH SCUDAMORE, daughter to Griffith Hamden of Hamden, in the county of Bucks, Esq.; first married to Edw. Oglethorpe, Esq., sonn & heir to Owen Oglethorpe, in the county of Oxford, Knight, and by him had 2 daughters; after to Sr. Phillip Scudamore of Burnham, in the county of Bucks, Kt., and lastly to Henry Leigh, Esq[r]., sonn and heir to Sr. Edw. Leigh, of Rushall, in the county of Stafford, Kt.; by him had one son, named Samuel, now living. She dyed at Croydon, March 28, 1649, being the 73rd year of her age.

Under the figure of a man in a gown, on a brass plate—

Sub hoc marmore sepultus est GULIELMUS MILL, generosus, duarum uxorum maritus; quarum prior erat Avisia, filia Edmundi Harwell de Besford in Wigorniensi agro, armigeri; e qua sustulit filios quatuor, Nicolaum, Gulielmum, Joannem, Thomam; filias quinque,

stated, from the Herald's Visitations, that Thomas Heron died in 1518; but the above date of 1544 is most probably correct.

Annam, Elizabetham, Franciscam, Milicentam, Margaretam. Posterior Margareta filia Nicolai Clerke de Ecleston in Eboracensi agro, generosi; e qua sustulit unicam filiam Margaretam. Septuagesimo ætatis anno mortuus est, Januarii duodecimo, anno Domini millesimo quingentesimo sexagesimo octavo.

On a brass plate, on a grave-stone—

Here under lieth the body of
Edward Arnold, a brewer of
this towne, about the age of 64
yeares, who deceased on the 10 day of
August, anno Dni. 1628.

St. Mary's Chauntry and North Aisle.

On a black marble ledger—

Here lyeth interred the body of Ann, the wife of Mr. William Wharam, citizen and shipwright of London, who departed this life the 4th of October, 1716, aged 52 years.

On a like ledger, with arms—in a chief, a demi-lion and cross croslet, empaling *erm.* in a chief three lions ramp.; crest, a demi-lion ramp.—

Here lyeth the bodies of Mr.
William and Mary, son and
daughter of Mr. William Boddington,
of London, and Frances his wife.
Mary departed this life the 13th of
July, 1695, aged 14 yeares; William
(Cursiter) departed the 25th of
November, 1703, aged 26 years.
William the father departed the 10th of
February, 1718, aged 73.
Frances the mother departed the 11th of
November, 1727, aged 84.

On a brass plate—

Here lyeth the bodyes of Robert
Jackson, yeoman, the sonne of Nicholas
Jackson, and Anne his wife, daughter of
Richard Wood, yeoman, who had issue by
her 9 children, whereof 3 were living at
his decease, the 21 daye of September, 1622;
and Anne his wife died the 30 of August, 1612.

On a brass plate, beneath the figures of a man and woman—

Here lyeth buried the body of Robert
Jackson the younger, yeoman. He married
Elizabeth Wackrell, daughter of Richard
Wackrell, yeoman, who departed this life
xith of October, anno Dom. 1629.
For whose pious memorie his loving
wife caused this memoriall.
They had issue 17 children, 12 sonnes and 5 daughters.

On a black marble ledger—

Here lyeth the body of Francis Fletcher, late citizen of London, and a friend to mankind, who departed this life the 4th of July, 1757, aged 58.

Also the remains of Mary Fletcher, relict of Francis aforesaid; she departed this life April 11, 1771, aged 65 years.

On a Portland stone ledger—

Ann Lodge, daughter of James and Mary Lodge, departed this life June 29, 1772, aged two years.

On a black marble ledger—

Here lyeth the body of Benjamin Bowles, Esq., who died October 6, 1776, in the 60th year of his age.

St. Nicholas' Chauntry, and South Aisle.

On a black marble ledger—

Depositum
GULIELMI WAKE,
Archiepiscopi Cantuariensis,
Qui obiit XXIV Januarii, anno Dom. MDCCXXXVI.
Ætatis suæ LXXIX.
Et
ETHELDREDÆ, uxoris ejus,
Quæ obiit XI Aprilis, MDCCXXXV,
Ætatis suæ LXII.

On a Portland stone slack, with marble ledger—

Here lieth the body of
the Most Reverend
JOHN POTTER,
Archbishop of CANTERBURY;
who died
October X, MDCCXLVII,
in the LXXIVth year of his age*.

On a black marble ledger—

Here lieth the body of WILLIAM HERRING, Esqr., who died XXVIII Sep. MDCCCI, in the LXXXII year of his age.

On a like ledger—

Here lieth the body of MONTAGUE DOROTHY, wife of WILLIAM HERRING, Esq., who died XXVII July, MDCCLXXXIX, in the LXV year of her age.

* This ledger being concealed by the erection of pews in this part of the church, the marble tablet before mentioned was placed on the wall.

On a marble tomb, at the foot of Murgatroid's—

Here lieth ELIZABETH BRADBURY,
Wyfe unto Wymond Bradbury, of
Newport-pond, in Essex, Gent., daughter
to William Whitgifte, of Claveringe, in
the county aforesaide, Gent., and second
brother to Doctor John Whitgifte, Arch-
bishoppe of Canterbury; and who had
issue by her abovenamed husband, Jane,
William, Anne, and Thomas; and deceased
the 26 day of June, an. Dni., 1612, being
of the age of 38 yeares and three
months.

On a brass plate, beneath the figures of a man in armour, and a woman kneeling—

Here lyeth the bodies of THOMAS WALSHE, of Croydon, gentleman, 3rd Sonn of Fraunces Walshe, of Sheldisley Walshe, inn the County of Worcester, Esquier, and KATHERIN, his Wife, Daughter of William Butler, of Tyes in Sussex, Gent., whoe had by her too sonnes and one daughter, viz. Fraunces, John, & Avice; which John died younge, and the aforesaide Thomas Walshe departed the xxx of August, 1600.

Arms—quarterly 1 and 4 *arg.*, a fesse betw. six martlets *sa.*, 2 and 3 *arg.*, a chev. betw. three roses *gu.*, with an annulet for difference, empaling . . . . . . *.

On a black marble ledger, by Archbishop Whitgift's tomb—

* These arms are restored from the MS. cited in p. 199.

M. S.
To the Memorie of y^t Worthy
Lady Elizabeth Gresham,
Late wife of Sir William
Gresham, Knight, who, after
she had lived 72 yeares,
Unspotted in her Conversation,
Charitable to the poore,
Sincere in Religion, Re-
signed up her soule into the
hands of her Creator, upon y^e
9 day of December, 1632,
& Lieth here interred in
hope of a glorious
Resurrection.
For a Memoriall of which
Singular virtues her deare &
only Daughter, E. G., hath
consecrated this marble as a
Duty she could
Performe.

On a Portland stone ledger—

Here lieth the body of John Usborn,
citizen and stationer of London,
who died the 3rd of November, 1738, aged 70 years.
He was remarkable for his piety to God,
and his benevolence to his fellow creatures.
Also,
Grace, the wife of the said John Usborn,
who died July the 17th, in the 76th year of her age.

On a brass plate, under the indents of some figures—

Here under lyeth the body of John Davenant, Citizen and Marchant Taylor of London, who had to wyfe Margaret Clarke, and had issue by her x sonnes and iiij daughters. He, being about the age of lx and one yeres, deceased the xxiijth of October, Anno Domini, 1596.

Arms—quarterly 1 and 4 *gu.*, semèe of cross croslets fitchey *or*, three escallops *erm.*; 2 and 3 vairè *gu.* and *sa.*, a canton of the first, empaling *or;* a fesse between three lions couchant regardant *gu.**

On a brass plate, on a rough black stone—

RICHARD YEOMAN, farmer, of Waddon
Courte, the husband of 3 wives, by
whom he had 9 children; 5 by the
first, 2 by the nexte, and 2 by the laste:
and deceasing the xxvith daye of
December, here lyeth buried,
Anno Dni. 1590, ætatis suæ, 90.

On a brass plate, on a white stone—

Here lieth the body of THOMAS YEOMANS,
who had issue, by Anne his wife, George
and Susan; which Thomas deceased the
first of Aprill, An. Dni. 1602.

On a Portland stone ledger—

WILLIAM CHAPMAN of Doctors' Commons, Gent., died December 23, 1730, aged 52 years.

Cross Aisle.

Between the middle and south aisle, on a brass plate, on a greyish stone—

Under this stone lyeth interred the body of John Woode, late of Croydon, In-holder, who had 2 wyfes, Anne and Amy; by his first wife he had allone yssueie 7 sonnes; and by the last, 3 sonnes and 4

* Restored from the before-mentioned paper.

daughters. He deceased the 23 day of June, beinge Saterday, an. Dni, 1525, ætatis sue 52.

On the same stone, on another brass plate—

Olim et talis ego qualis nunc tu esse videris,
Olim et tu talis nunc ego qualis eris.
Terra tegit cineres, humus est aptata sepulchro,
Quid nisi pulvis humo terra sepulta cinis.

On a small free-stone—

Here lyeth the body of WILLIAM MICHELL, who departed this life the 17 of July, 1658, aged 60 years.

On a Portland stone ledger—

Captain GEORGE PROTHEROE departed this life the 25th of February, 1745, aged 70.

On a like stone—

The Rev. Mr. JAMES GARDNER, rector of Slingsby, in the county of York, died December 11, 1772, in his 88th year*.

EPITAPHS IN THE CHURCH YARD.

Of the many inscriptions in the church-yard, the following only are worthy of record.

The following inscription, preserved by Aubrey, was recovered from a MS. once belonging to Augustine Vincent, Esq., Windsor Herald—

* There are several inscriptions in this aisle, and particularly in the belfry, which I could not decypher, as also one or two in other parts. The illegibility of those in the belfry are caused, in a great measure, by the feet of the bell-ringers.

Here lyeth John Redynge, Esq., late Treasurer to Prince Henry, sonne to King Henry viii. and Marye his wife, Mistress to the Prince of Cassel; which John deceased the xix daye of January, anno D'ni mccccclxxx*.

On a free-stone tomb, supported with brick, near the north entrance—

Mortis Trophæum
de corpore HENRICI HOAR, Medico-Chirurgi,
qui prisci candoris et humanitatis
se exemplum præbuit,
et plane bonus fuit licet optimis comparetur.
XI Februarii obiit, anno salutis MDCCIX, ætatis LXXII.
Annis ille senex fuit et candore;
sed illum dixerunt omnes non satis esse senem.

On a stone—

Here lies the body of ANNA, the loving and beloved wife of ROGER ANDERSON, of London, youngest of the seven sons of William and Bridget Anderson, of this parish. She was daughter of the Rev. Dr. Casson, rector of Sutton, in Herefordshire, and one of the prebends of Hereford Minster; a great sufferer during the time of Cromwell's usurpation, for his firm adherence to the Church of England, and his loyalty to the royal martyr. She died 19 Jan. 1723, in the 74th year of her age.

Finis coronat opus.

On a large vault, bricked high above ground, and covered, near the north entrance to the church yard—

This is the burial place of the GARDINER's family, of Haling.

* I have not been able to find any other mention of this gentleman, whose tomb was so singularly inscribed; but in the register, among the burials, is entered the name of one "Wyllm Readyng," who died January 7, 1562.

Outside of the wall of St. Nicholas' Chauntry, on a white marble tomb, inclosed with iron palisades, is this inscription:—

Beneath this Tomb repose the remains of
The Right Hon. Lady CATHARINE SHELDON, late PHIPPS,
who died in January, 1738;
JOHN SHELDON, Esq., of Mitcham,
who died in March, 1752;
The Right Hon. CONSTANTINE PHIPPS, Baron Mulgrave,
who died in September, 1775;
The Right Hon. Lady LEPEL PHIPPS, Baroness Mulgrave,
who died in March, 1780;
RICHARD SHELDON, Esq., of Lincoln's-inn-fields, who died
the 15th February, 1795, aged 72 years.

On a grave-stone—

Here lyeth the body of
JOHN D'ARLEY,
son of Major D'Arley,
who departed this life
on the 23rd day of November, 1828,
aged 17 years.

The following inscriptions are now lost:—

In memory of URSULA SWINBOURN,
who,
after fulfilling her duty
in that station of life her Creator had allotted her,
and by her faithful and affectionate conduct,
in a series of thirty-five years,
rendered herself respected and beloved whilst living,
and her loss sincerely regretted by the family she lived with,
departed this life the 5th of January, 1781, aged 55.
Reader,
Let not a fancied inferiority,
from her station in life,

prevent thy regarding her example;
but remember,
according to the number of talents given,
shall the increase be expected.

Mr. William Burnett, born January 29, 1685;
died October the 29th, 1760.

What is man?—
To-day he's drest in gold and silver bright,
Wrapt in a shroud before to-morrow night;
To-day he's feasting on delicious food,
To-morrow, nothing eats can do him good;
To-day he's nice, and scorns to feed on crumbs,
In a few days, himself a dish for worms;
To-day he's honour'd, and in great esteem,
To-morrow, not a beggar values him;
To-day he rises from a velvet bed,
To-morrow lies in one that's made of lead;
To-day, his house, though large, he thinks too small,
To-morrow, can command no house at all;
To-day, has twenty servants at his gate,
To-morrow, scarcely one will deign to wait;
To-day, perfum'd, and sweet as is the rose,
To-morrow, stinks in every body's nose;
To-day he's grand, majestic, all delight,
Ghastly and pale before to-morrow night.
Now, when you've wrote and said whate'er you can,
This is the best that you can say of man!

CHAPTER VIII.

Benefactions to Croydon.

The following list, shewing the different benefactions to this town, is copied from a table in the church:—

ESTATES AND CHARITIES BELONGING TO THE PARISH OF CROYDON.

1528. A rent-charge, given to the poor of the Little Alms House, by Joan Price, of £1 *per annum*.

1566. A piece of ground whereon the great market-house is built, and towards the building whereof Francis Tyrrell, citizen and grocer of London, gave 200*l*.

1614. Seven acres of land, near Hermitage-lane, the gift of Edward Croft.

1619. A tenement or stable, called the Old Shop, near the Butcher Row.

Two tenements by the Mint Walk, one called Parkhurst.

1622. A farm house and 100 acres of land, called Stockbenden, at Limpsfield, in Surry, chiefly the gift of Henry Smith, Esq.

1624. Twenty-six acres of land at New Cross, near Deptford, whereon have been lately erected a number of dwelling houses, the gift of the said Henry Smith.

1627. The ground whereon the workhouse stands, the gift of Sir Wm. Walter.

1629. The little Alms-house, for the maintenance of eight poor people of this parish; towards the rebuilding whereof Arnold Goldwell gave 40*l.*

16 The Fishmongers' Company in London pay the said Alms-house 2*l.* 13*s.* 4*d.* *per annum**.

1708. The Butter Market, rebuilt by Archbishop Tennison.

A rent-charge, given by Mr. Rowland Kilner to the Little Alms-house, of 5*l.* *per annum.*

1760. 155*l.*, the gift of Mr. Joseph Williams and others, the interest whereof to be annually disposed of in bread for the poor.

Incorporated, with Governors and Trustees.

1443. The great Alms-house, for a tutor and six poor people, the gift of Ellis Davy, citizen and mercer; with land and tenements for their maintenance, the gift of Ellis Davy, citizen and mercer of London.

1599. The Hospital of the Holy Trinity, for a schoolmaster, warden, and twenty-eight poor men and women of Croydon and Lambeth, with lands and tenements for their maintenance, the gift of Archbishop Whitgift.

1619. Two messuages or tenements in Northampton, given to the said hospital by the Rev. Dr. Prethergh.

A rent-charge, payable out of a tenement in St. Paul's Churchyard, London, given to the said hospital by Mr. Edward Barker, amounting to 6*l.* 13*s.* 4*d.*

3-5ths of a farm at Mitcham in Surry, given to the said hospital by Ralph Snow, Esq.

A tenement and piece of ground in the Butcher Row, given to the said hospital by Mr. Richard Stockdale.

A dinner, yearly, to the said hospital, for which the Fishmongers' Company in London pay 13*s.* 4*d.*, and put into the box 10*s.*†

* Left by Lady Ann Allot, of Sunderstead, circa 1600.

† Left by William Barlow, D.D., Bishop of Lincoln, who died in 1690. He also bequeathed 13*s.* 4*d.* *per annum* to a licensed preacher,

A farm at Horn, in Surry, for binding poor boys apprentice, the gift of Archbishop Laud*.

1714. A school-house at Croydon, and two farms at Limpsfield, for educating 10 boys and 10 girls; the gift of Archbishop Tenison.

1760. 195*l.*, the gift of the Earl of Bristol and several inhabitants of Croydon, for enlarging the little Alms-house.

1*l.* 10*s.*, the annual gift of ——— Wigsell, Esq., Sanderstead, to be given in bread to the poor widows, on St. Thomas' day.

1790. 163*l.* 13*s.* 6*d.* 3 *per cts.*; the interest of which to be applied annually to poor families of this parish; the gift of the Rev. East Apthorp, D.D., late vicar of this parish.

On another Table, in the Church.

1831. 100*l.*, the gift of Mrs. Mary Allen, of Camberwell; the interest whereof is to be annually disposed of in bread for the poor of this parish, at Christmas-day, for ever.

Other Benefactions to the Town of Croydon.

Archbishop Parker bequeathed to the poor of Croydon and Lambeth, 30*l.*—Archbishop Grin-

to preach a sermon in Croydon Church, on the 22nd March, yearly, being the day on which the hospital of Archbishop Whitgift was founded; and to the vicar of Croydon 3*s.* 4*d.* for giving notice of the sermon on the preceding Sunday. The money to be paid by the Fishmongers' Company.

* In 1635 Archbishop Laud purchased, for 300*l.*, a messuage and lands at Albury, Warwickshire, in the name of Sir John Tonstall and others, the rent of which estate to be applied to the apprenticing poor children of this parish. The estate was afterwards sold, pursuant to a decree of the Court of Chancery, in 1656, for 225*l.*;

dall bequeathed, to purchase lands for the benefit of the poor of the Little Alms-house, 50*l.*; with which sum the vicar and churchwardens for the time being purchased a copyhold house in Waddon, 11th November, 1583. Also, to the poor of Croydon and Lambeth, 20*l.*—Archbishop Whitgift left, by will, to the poor of Croydon, 20*l.*—Archbishop Bancroft, ditto, 20*l.*—Archbishop Abbot, ditto, 20*l.*—Archbishop Laud, ditto, 10*l.*—Archbishop Juxon, ditto, 100*l.*—Archbishop Sheldon, ditto, 40*l.*—Archbishop Tenison, ditto, 40*l.*; also, to Whitgift's Hospital, 100*l.*—Archbishop Wake, for binding out apprentices, 40*l.*—Archbishop Secker, for the poor of Croydon, 500*l.* 3 *per cent.* Consols.

Jasper Yardley, Gent., who died 31st May, 1639, and lies buried in Guildford Church, gave, as appears from the inscription on his tomb, to the parishes of Croydon, St. George's, Southwark, and Lambeth, 40*l.* each, "to be put fourth yearly, gratis, by their churchwardens, in 5*l.* parcels, to the severel poor of each parish, for stock to set them to work, or for tradinge. He also gave legacies to all the poore of the Hospitallis of Guildford and Croydon."

which sum, augmented by the addition of 35*l.*, raised by the trustees among themselves, purchased the above-named farm.

APPENDIX.

No. 1.

Instrumentum factum super Appropriatione Ecclie de Croydon & Assignatione Manerii de Woddon, dat' 16 *Jan'* 1390. *Reg. Courtnay, fol.* 179. *b.*

In nomine Dñi, Amen. Anno ab incarnacõe ejusdem secundum cursum & computacõem ecclie Anglican' millmo ccc^{mo} nonagesimo, indicõe xiiii^{a}, pontificatus sanctissimi in Christo pris & dni nri dni Bonifacii divina providen' pape Noni anno secundo, mensis Januarii die sexto decimo, in ecclia pochiali de Croydon, Wintonien' jurisdicõis immediate reverendissimi pris dni archiepi Cantuarien' in nror' notarior' & testium subscriptor' presencia coram venli in Xpo pre ac dno dno Roberto Dei gra Londonien' epo in causa sive negocio unionis annexionis sive incorporacõis dce ecclie pochialis de Croydon ad collacõem reverendissimi pris dni archiepi Cant' notorie ptinentis prioratui de Bermondsey ordinis Cluniacen' ac concessionis grangie ejusdem prioratus manerium de Woddon nuncupate mense arshiep̃ali Cantuar' judice uno delegato juxta formam subscriptam a p̃dc̃o d̃no ñro pape sufficient' & legitime deputato in p̃dc̃a eccl̃ia ad infrascripta faciend' judicialit' sedente, reverendissimus in Christo pater & d̃nus d̃nus Willm̃us Courtenay Dei gr̃a archiep̃us Cantuarien' tocius Anglie primas & ap̃lice sedis legatus, & religiosus vir frat' Johannes prior p̃dc̃i p̃oratus de Bermondesey personalit' ꝑ seipsis & nominib' ꝓpriis, ac discreti viri mag' Walterus Gybbes licentiatus in legibus & frat' Thomas Fakenham monachus p̃fati p̃oratus ꝓcuratores ejusdem prioris & conventus de Bermondesey nom̃ibus eor̃dem d̃nor' suor' judicialit' comparentes, quasdam l̃ras ap̃lcas predc̃i d̃ni ñri pape sanas & integras, omni prorsus vicio & sinistra suspic̃õe carentes, more Roman' cur' cum filis sericis & sigillo plumbeo bullatas, concernentes unionem, annexionem, sive incorporac̃õem eccl̃ie, ac concessionem & assignac̃õem grangie sive manerii p̃dc̃or' eidem Dño Londoniens' Ep̃o judicii delegato, una cum ꝓcurator' d̃cor' ꝓcurator' sigillo eor' con' judicialit' p̃sentarunt, quas quidem l̃ras ap̃licas & quod quidem ꝓcuratoriũ idem reverendus pat' d̃nus ep̃us delegatus reverent' recepit, & illas ac illud corã eo publice legi fecit, quor' tenores in hæc verba secuntur.

Tenor Bullæ. "BONIFACIUS Ep̃us servus servor' Dei venerabili ep̃o Londonien' sal̃t' & ap̃lican ben'. Sincere devoc̃õis affectus quem venerabilis frater noster Will̃mus archiep̃us Cantuarien' & dilecti filii prior & conventus de Bermondesey Cluniacen' ordinis Wintonien' dioc' ad nos & Roman' gerunt eccl̃iam promeretur ut votis eor' illis presertim ꝑ que mense archiep̃alis Cantuarien' & dci prioratus utilitas procuratur favorabilit' annuamus. Exhibita siquidem nobis nuꝑ ꝑ pte archiep̃i ac prioris & conventus p̃dc̃or' peticio continebat quod iidem prior & conventus habent quoddam manerium sive quandam grangiam ad manerium de Woddon nuncupatũ cuidam alteri manerio prefate dioc' ad dctam mensã ꝑtinent' contiguum, ac ꝑochialis eccl̃ia de Croydon ejusdem dioc' & collac̃oem archiep̃i Cantuarien' pro tempore existentis pertinet, quodq; si eccl̃ia prioratui uniretur & manerium de Woddon, predicte mense prefatis concederetur & assignaretur, in utilitatem & commodum prioratus & mense cederet eor̃dem. Quare ꝑ pte dictor' archiep̃i ac prioris & conventus nobis fuit humiliter supplicatum ut alicui prelato in partibus illis committere & mandare dignaremur quod de p̃missis diligenter se informet, & si ꝑ informac̃oem hujusmodi reperiret quod unio & connexio hujusmodi si fierent in utilitatem & commodum prioratus & mense p̃dc̃or' cederent, auctoritate ñre eccl̃iam prioratui uniret, annecteret, & incorporaret, ac mañium de Woddon p̃dc̃e mense p̃fatis de consensu & voluntate dcor' prioris & conventus, eciam absq; licentia vel consensu abbatis Cluniacen' & prioris de Caritate ꝑ priorem soliti gubernari dicti ordinis Matisconenem & Antisiodern' dioc' pro tempore existentiũ & conventũ monasterior' a quo quidem monasterio de caritate pfat' prioratus dependet, concederet & assignaret. Nos igitur hujusmodi supplicationibus inclinati fraternitati tue de qua in hiis & aliis spial̃em in dño fiduciam obtinemus ꝑ ap̃lica scripta committimus & mandamus quatenus suꝑ p̃missis auctoritate ñra te diligenter informes, & si ꝑ informac̃oem hujusmodi ita esse reperis prenominatam eccliam cujus centum cum omnibus juribus & p̃tinenciis suis eidem p̃oratui cujus octingentar' marcar' sterlingor' fructus redditus & proventus secundũ communem estimacõem valorem annuum ut ipsi prior & conventus dci p̃oratus asseruerunt non exedunt, auctoritate p̃dc̃a ppetuo unias, incorpores, & annectas. Ita quod cedente vel decedente rectore dce ecclie qui nunc est, vel eccliam ipsam alias quomodolt dimittente, liceat ipsis p̃ori & conventui p̃oratus possessionem dce ecclie apprehendere & ex tunc ppetuo retinere, alicujus licentia sup hoc minime requisita. Reservatã tamen de fructib9, redditib9, & pventib9 ejusdem eccle p ppetuo vicario in ea instituendo si illa que ab olim reservata fuisse dicitur congrua non sit congrua porcõne de qua idem vicarius possit congrue sustentari ep̃alia jura solvere & alia incumbencia sibi onera supportare, cujus quidem porcõis ab olim ut p̃fertur reservate collacio ad rectorem dce ecclie pro tempore existentem ptinet, & quam seu illam que p te vigore p̃sentiũ reservetur si postquam unio, annexio, & incorporacio hujusmodi effectum sortiti fuerint ad dc̃i archiep̃i ejusq; successor archiep̃or' Cantuarien' quietant p tempore collacõem ppetuo volumus ptinere. Et nichilominus p̃fatum manerium de Woddon cum omnibus juribus & ptinenciis suis p̃dc̃e mense de consensu tamen & voluntate dctor' p̃oris & conventus ipsius p̃oratus eciam absq; licentia abbis Cluniacen' & p̃oris qui sunt p tempore & conventus de Caritate monasterior' p̃fator'. Cum ille qui Cluniacen' & ille qui de Caritate monasteriis degunt ad presens iniquitatis alumpno Roberto olim Basilice duodecim apostolor' de urbe presbytero cardinali nunc antipape qui se Clementem septimum ausu sacrilego nominare presumit nator' adherere & favere presumant eadem auctoritate concedere & assignare ꝓcures. Inducens p te vel alium seu alios p̃dc̃um archiep̃um vel ꝓcuratorem suum ejus nomine in corporalem possessionem mañii de Woddon ac ipsius jurium & ꝑtinenciar' p̃dict' & defendas inductum sibique faciens de ipsius manerii de Woddon fructibus, redditibus, ꝓventibus, juribus, & obven©õnibus universis integre responderi; contradictores p censuram eccĩasticam appella©õne postposita compescendo. Non obstantibus si aliqui sup ꝓvisionibus sibi faciendis de ꝓchialibus eccłiis vel aliis beneficiis eccłiasticis in illis ꝑtibus speciales vel generales ap̃lice sedis vel legator' ejus łras impetrarint etiam si ꝑ eas ad inhibi©õem, reserva©õem, & decretum vel alias quomodołt sit processum, quos quidẽ łras & ꝓcessus ear' vigore habitos vel habendos ad p̃fatã eccłiam volumus non extendi sed nullum p hoc eis quoad assecu©õem ꝓchialium eccłiar' & beneficior' alior' prejudicium generari seu si aliquibus conconjunctim vel divisim a dicta sit sede indultum quod interdici, suspendi, vel excommunicari non possint p literas ap̃licas non facientes plenam & expressam & de verbo ad verbum de indulto hujus men©õem & quibuslibet aliis privilegiis & indulgenciis ac łris ap̃licis generalibus vel specialibus quor̃cunq, tenor' existant p que presentibus non expressa vel totaliter non incerta effectus eor' impediri valeat quomodolib' vel differri & de quibus quor̃q, totis tenoribus de verbo ad verbum habenda sit ñris literis mentio sp̃ialis. Nos enim ex nunc irritum discernimus & inane si secus super his a quoquam quavis auctoritate scienter vel ignoranter contigit attemptari. Dat' Rome apud sanctum Petrum quinto cal' Octobr', Pontificatus n̄ri anno primo. Pateat universis (procuratoriū prioris & conventus de Bermondesey, qui nominant Mag' Walterum Gybbes in legibus licentiatum, & d̄num Wilt̄mum Baunton rect' eccl̄ie de Harewe, & fratrem Thomam Fakenham monachum confratrem suum, conjunctim & divisim, legitimos ꝑcuratores suos ad comparendum coram Roberto ep̄o Lond' & supra dc̄as l̄ras ap̄licas de unione eccl̄ie de Croydon conventui supradc̄o facienda, & manerio de Woddon mense archiep̄ali assignando d̄cto d̄no p̄sentendas & ad omnia alia ꝑficienda quæ dc̄um negocium spectant', dat' in capitulari 16 Jan. 1390), & in hujusmodi procuratorio tenor dc̄e bulle totalit' inseritur ante dat' ejusdem ꝑcuratorii. Post quor' quidem l̄rar' apl̄icar' & ꝑcuratorii p̄dc̄or' recepc̄ōem & lecturam, p̄dc̄us frater Johannes prior prioratus de Bermondesey cor' p̄fato d̄no ep̄o judice delegato palam & expresse ꝑtestabatur ac ꝑtestatur quod ꝑ aliquā ipsius in causa sive negocio p̄dct' coram eodem d̄no judice delegato comparic̄ōem ꝑsonalem noluit nec intendebat, non vult nec intendit, potestatem p̄dc̄or' ꝑcurator' suor' in aliquo revocare, & subsequent' incontinenti memoratus d̄nus archiep̄us & p̄fati Mag' Walterus & frater Thomas ꝑcuratores nōibus dc̄or' d̄nor' suor' p̄dc̄um d̄num ep̄um judicem delegatum cum instantia requisiverunt, & eor' quil̄t requisivit, quatenus idem d̄nus ep̄us judex delegatus in causa sive negocio supradc̄o juxta omnem, vim, formam, et effectum d̄ctar' l̄rar' ap̄licar' procedere, et ea omnia et sinḡla que sibi in eisdem l̄ris ap̄licis exequenda et facienda committuntur exequi et facere dignaretur, unde p̄dc̄us d̄nus ep̄us judex delegatus volens ut asseruit mandatis ap̄licis humilit' obedire villam de

Croydon Wintonien' dioc' in cujus pochiali ecctia inibi consistente ad infrascripta expedienda judicialit' sedebat et sedet in quem locum p̃dict' d̃nus archiep̃us et mag' Walterus ac frater Thomas ꝑcuratores nominibus dcor' d̃nor' suor' expresse consencierunt et quilt eor' consenciit eo quod idem locus est ex vulgi opinione insignis communiter reputatus, ac ex aliis certis et legitimis de causis ad ejus animum ut asseruit moventibus ad dctor' d̃ni archiep̃i ac mag̃ri Walteri et f̃ris Thomæ ꝑcurator' petic̃ōem in hac parte pronunciavit et ꝑnunciat insignẽ et pro informac̃ōe dc̃i d̃ni ep̃i judicis delegati in hac pte habenda p̃dc̃us d̃nus archiep̃us ꝑ se et nomine ꝑprio ac p̃scripti mag̃r Walterus et frater Thomas ꝑcuratores nomine quo supra coram eodem d̃no ep̃o judice delegato quam plures testes fide dignos ad probandum suggesta et deducta in dc̃is lris ap̃licis fuisse et esse vera judicialit' pduxerunt et pducunt. Quos quidem testes sepe dictus d̃nus ep̃us judex delegatus ad ipsor' d̃ni archiep̃i et ꝑcurator' p̃dc̃or' petic̃ōem admisit et admittit quibus quidem testibus ad sancta Dei Evangelia p eos corporalit' coram ipso d̃no delegato et de ipsius mandato tacta juratis idem d̃nus judex delegatus eosdem testes sup c̃tis articulis a tenore p̃dc̃ar' ap̃licar' extractis diligent' et singillatim examinavit, ac eor' dc̃a et deposiciones in scriptas p eos notarios subscriptos in causa sive negocio hujusmodi coram eo actorum scribas redigi fecit, et deinde memoratus d̃nus archiep̃us et p̃fati mag̃r Walterus ac frat' Thomas ꝑcuratores in nomine ꝑcuratorio quo supra quandam ordinac̃ōem congrue porcionis vicarie dc̃e ecctie de Croydon ab antiquo dotate factam p recolende memorie d̃num Johannem Stratford quondam archiep̃um Cantuarien' ipsius sigillo roboratam corã p̃dc̃o d̃no ep̃o judice delegato judicialit' exhibuerunt et quit̃ eor' exhibuit tenorem qui sequitur continentem. Johannes permissione divina Cantuarien' tocius Anglie primas et ap̃lice sedis legatus cunctis Christi fidelibus salt' in d̃no consequi sempiternam; ex nostro mero pastorali officio nup subditos ñros d̃num Joh̃em de Tonneford rectorem poch' eccl̃ie de Croydon et d̃num Joh̃em de Horstede ppetuum vicariũ ejusdem ñre jurisdictionis immediate ac p̃emptor' exhibend' ordinac̃ōem porc̃ōem dict' vicarie si quam haberent certis die et loco competentibus citari fecimus coram nobis, qui juxta vocationem hujusmodi corã ñro in ea pte commissario in judicio comparentes asseruerunt se nullam ipsius vicarie ordinac̃ōem habere, petieruntque a nobis instanter ut dc̃am vicariam faceremus in certis et indubitatis dc̃e eccl̃ie porc̃ōibus ordinari, unde nos tam sup vero valore annuo fructuum reddituum et pventuum dc̃e ecc̃ie quos idem rector pcipit et pcipere consuevit quam super vero valore annuo alior' pventuum, oblac̃ōnum et obvenc̃ōnum ejusdem eccl̃ie p ipsius vicarium hactenus pceptor' ac eciam de et sup omnibus et sing̃lis oneribus p̃fate eccl̃ie incumbentibus p pochianos dc̃e eccl̃e plenam in ea pte noticiam obtinentes primitus inquiri fecimus, et deinde de consensu rectoris et vicarii p̃dc̃or' porc̃ōem vicarie dc̃e eccl̃ie consideratis ipsius facultatibus et oneribus ipsi eccl̃ie incumbentibus ac ceteris in ea pte undique ponderandis, modo infra scripto ordinandam duximus et taxandam ac eciam declarandã quid et quantum p̃fatus vicarius et successores sui pciperent de fructibus, redditib' et pventibus eccl̃ie memorate, necnon que onera d̃ctis rectori et vicario ac

Ordinatio porcionis vicarii ab olim facta.

successoribus suis incumbere debeant in futur'. Ordinam' siquidem et statuimus quod rector dc̃e eccl̃ie qui ꝑ tempore fuerit omnes decimas majores, viz. Blador' feni, sylvæ cedue, et lignor' arbor' ceduar' excisar' infra fines et limites dc̃e ꝑochie ꝑvenientes ac eciam omnia mortuaria viva occasione sepulture cujuscunq' ad dc̃am eccl̃iam obventura et spectantia et spectare valencia seu debencia qualitercunq' medietatemq' decimar' agnor' decimabilium qui ꝑ capita decimari debent de consuetudine vel de jure infra ꝑochiam eccl̃ie antedc̃e ꝓvenienciũ, nec non pensionem octo marcar' ꝑ equal' porc̃ões in festis Sc̃i Michaelis natal' d̃ni pasch' et nativitatis Sc̃i Joh̃is Baptiste ꝑcipiend' annis singl̃is a vicario dc̃e eccl̃ie qui ꝑ tempore fuerit, et vicaria p̃dc̃a nec non reddit' fructus et p̃ventus dc̃e eccl̃ie jura, commoditates et quascunq' res alias ad dc̃am ꝑtinentes seu spectare debentes eccl̃iam vicario dc̃e eccl̃ie inferius non ascripta ꝑcipiat et habeat in futur'. Item q̃d dc̃us vicarius et successores sui ib̃m vicarii habeant et teneant mansum solitum dc̃e vicarie cum gardino adjacente eidem. Item habeant et ꝑcipiant vicarii dc̃e eccl̃ie nomine vicarii p̃dc̃i omnes et omnimodas oblaciones in dc̃a eccl̃ia de Croydon et in quibuscunq' locis infra fines, limites, seu decimac̃ões ejusdem situatas qualitercunq' factas et faciendas seu ad eam vel in ea ꝓvenientes et imposter' ꝓvenire valentes modo, causa, occasione, vel colore eciam quibuscunq'. Item habeant et ꝑcipiant dc̃i vicarii nomine dc̃e vicarie medietatem decimar' agnor' decimabilium ꝑ capita ut p̃mittitur decimandar' ꝑ vicarium dc̃e eccl̃ie colligendor' et denarios ꝓvenientes ex vel ꝑ illis qui non fuerint ꝑ capita decimati de consuetudine vel de jure, nec non et omnes decimas lane, vitulor' porcellor' aucar' anatum, columbarum, casei, lactis, lacticinii, herbagii, pomor' piror' et alior' fructuum in gardinis et ortis crescentiū eciam pedefossator' nec non lini, canapis, ovor' mercimonior' et molendinor' omnibus infra fines et limites seu decimationes pochie dc̃e eccłie jam constructor' et imposter' construendor' omnesq' alias minutas decimas p̃fato rectori non ascriptas qualitercunq' spectantes et ptinentes ad eccłiam antedc̃am, nec non quecunq' legata relicta imposter' dc̃e eccłie que ipsius rectores seu vicarii possent de jure vel consuetudine pcipere et habere, ac eciam mortuaria omnia mortua seu non viva occasione sepulture cujuscunq' ad dc̃am eccłiam obventura seu spectare debentia quovismodo pveniencia infra pochiam eccłie antedc̃e, singuli autem vicarii p̃fationes deserviendi p se et alium presbyterum ydoneum p̃fate eccłie in divinis onus eciam ministrac̃ōnis, panis, vini, luminar' et omnium ac singular' que ib̃m ad celebrac̃ōnem divinor' in rebus vel personis necessaria fuerint, nec non et onus invenc̃ōis seu exhibic̃ōis et reparac̃ōis libror' sup pellicior' vestimentor' et ornamentor' dc̃e eccłie que p eccłiar' rectores seu vicarios inveniri seu exhiberi vel repari de jure vel consuetudine debent aut solent, ac insup onus solutionis decimar' et aliar' imposic̃ōnum quarumcunq' que Anglican' eccłie qualitercunq' imponi continget p quemcunq' quavis occ̃ōe vel causa juxta taxacionem dc̃e vicarie que ad decem libras sterlingor' taxari dinoscitur, et quam per sic taxata haberi volumus et mandamus ordinamus suis sumptib' subeant et expensis, onera vero repacois et refeccionis cancelli dc̃e eccłie, viz. in tectis et muris intus et exterius ac eciam cetera onera ordinaria et extraordinaria eidem eccłie incumbencia p̃fatis vicariis non ascripta supius rector dc̃e eccl̃ie qui ꝑ tempore fuerit ppetuũ subeat et agnoscat; ordinamus insup q̃d dc̃us vicarius et successores sui ib̃m vicarii juramentum ad sancta Dei Evangelia tacta corporaliter p̃stent rectori qui est seu erit imposter' eccl̃ie memorate q̃d in p̃missis vel circa ea seu eorum aliquod fraudem seu dolum nullatenus adhibebunt p se, alios, vel alium publice vel occulte, et q̃d de hujusmodi porc̃õne rectoris nichil sibi penitus usurpabunt: reservamus insup' nobis et successoribus ñris dc̃am vicariam augmentandi et diminuendi si et quando nobis et eis expedire videbitur plenariam potestatem. Dat' apud Maydenston 2 idus Junii A.D. mill^{mo} ccc^{o} quadragesimo octavo, et ñre translac̃õis quinto decimo. Post cujus quidem ordinac̃õis exhibic̃õem venerabilis vir mag̃r Joh̃es Godewyk Legum doctor rector dc̃e eccl̃ie de Croydon perantea publice reputatus cor' p̃fato đno ep̃o judice delegato judicialiter comparens non vi nec metu ductus sed ex sua certa scientia et spontanea voluntate dixit et in judicio fatebatur q̃d postquam habuit dc̃am eccl̃iam de Croydon fuit et est pochial̃e de Clyve Rossen' dioc' jurisdiccionis immediate dc̃i đni archiep̃i auctoritate ordinaria canonice et pacifice assecutus, et ea ꝑpter possessionem dc̃e eccl̃e de Croydon vacuam coram eodem đno ep̃o judice delegato pure et sponte dimissit ac dimittit totaliter re et verbo; tandem vero p̃dc̃us đnus ep̃us judex delegat' in dc̃a causa sive negocio ulterius p̃cedens inspectis per eum ut dixit et diligenter recensitis dc̃is exhibitis et p̃ductis, ac sufficienti et diligenti informac̃õe ut eciam duxit p eam recepta et habita sup omnibus et sing̃lis in p̃fatis l̃ris ap̃licis expressas ad finalem dc̃e cause sive negocii expedic̃õem ad peticōem et de expresso consensu p̃dc̄or' d̄ni archiep̃i ac mag̃ri Walteri et f̃ris Thome ꝓcurator' cor' eodem d̄no ep̃o judice delegato in judicio comparenciū de concilio eciam juris peritor' qui sibi tunc temporis assistebant dc̄am pochialem eccl̄iam de Croydon cum omnibus suis juribus et ꝑtinenciis universis, salva p̃dc̄a congrua porc̄ōne ꝓ vicario ejusdem eccl̄ie p̃dc̄o p̃oratui auctoritate ap̃lica, eidem d̄no ep̃o judici delegato in hac ꝑte commissa ꝑ suam sententiā univit, annexit, et incorporavit, unit, annectit, et incorporat, ac p̃fatum maneriū de Woddon cū suis juribus et ꝑtinenciis universis p̃dc̄e mense archiepali assignavit, et assignat, ac porcionem vicarie infra scripte congruam fuisse et esse ꝓnunciavit et declaravit, et p̃terea in dc̄a causa sive negocio juxta vim, formam, et effectum l̃rar' ap̃licar' p̃dc̄ar' ꝓcessit et ꝓcedit in hunc modum.

In Dei nomine, amen. Auditis et intellectis meritis cause sive negocii unionis, annexionis, sive incorporac̄ōnis eccl̄ie p̃ochialis de Croydon Winton dioc' ad collac̄ōem d̄ni arch' Cantuar' ꝑtinentis p̃oratui de Bermondesey ordinis Cluniacen' ac concessionis et assignac̄ōis grangie ipsor' manerium de Woddon nuncupate mense archiep̃ali Cantuar' auctoritate ap̃lica faciend' que seu quod coram nobis Roberto Dei gr̃a Lond' ep̃o sanctissimi in Xp̃o p̃ris et d̄ni d̄ni ñri Bonifacii divina ꝓvidencia pape noni in causa sive negocio p̃dc̄o judice delegato aliquamdiu vertebatur et pendet indecis'. Quia ꝑ testes coram nobis ꝓduct' juratos et examinatos et alia legitima documenta invenimus maneriū sive grangiam de Woddon p̃dct' d̄ctor' p̃oris et conventus manerio de Croydon mense archiep̃ali Cantuar' dc̄e dioc' ꝑtinenti contiguum fuisse et esse, eccl̄iamq' ipsam de Croydon p̃dct' ad collac̃ōnem d̃ni archiep̃i Cant' p̃tinere, ac q̃d si eccl̃ia ipsa ꝑoch' uniretur p̃oratui supradc̃o, et manerium de Woddon supradc̃um p̃dc̃e mense archiep̃ali concederetur et assignaretur, in utilitatem et commodum p̃oratus et mense cederet eor̃dem quodq' fructus, redditus, et ꝓventus dc̃e eccl̃ie ꝑoch' de Croydon centum ac prioratus p̃dct' octingentar' marcar' sterl' valorem annuum secundum communem estimac̃ōe non excedunt, porc̃ōnemq' vicarie dc̃e eccl̃ie de Croydon ab antiquo dotate, de qua idem vicar' poterit sustentari ep̃alia jura solvere, et alia onera sibi incumbencia supportare congruam et sufficientem, ceteraq' omnia et singula in l̃ris ap̃licis nobis in hac ꝑte directis suggesta ad quas referimus vera fuisse et esse. Eapropter nos Robertus ep̃us judex unicus delegatus p̃dc̃us Deum p̃ oculis habentes, Xp̃i nomine primitus invocato, de consilio jurisperitor' nobis assidenciũ, de et suꝑ p̃missis plenius informati, prehabitis et observatis in hac ꝑte de jure vel consuetudine requisitis, p̃fatam eccl̃iam de Croydon dc̃e dioc' jam vacantem cum suis juribus et ꝑtinenciis universis eidem p̃oratui auctoritate ap̃lica nobis in hac ꝑte commissa ꝑpetuo unimus, annectimus, et incorporamus, ita quod liceat priori et conventui p̃fati p̃oratus corporalem possionem dc̃e eccl̃ie apprehendere et ꝑpetue retinere, ac insuꝑ p̃fatum maneriũ de Woddon cum omnibus juribus et ꝑtinenciis suis p̃dc̃e mense archiep̃ali Cant' de consensu et voluntate dc̃or' p̃oris et conventus expressis eadem auctoritate ap̃lica concedimus et assignamus, d̃cumq' reverendissimũ in Xp̃o patrem d̃num Will̃mum archiep̃um vel ꝓcuratorem suum ejus nõie in corporalem possessionem mañii de Woddon p̃dc̃i ac ipsius juriũ et p̃tinenciar' p̃dc̃or' inducendum fore decernimus, porcõnemq' dce vicarie de qua idem vicarius possit congrue sustentari, epalia jura solvere et alia incumbencia sibi onera supportare ab antiquo dotatam congruã sufficientem fuisse et esse pnunciamus et declaramus in hiis scriptis. Demum vero plecta pdcum dnum epum judicem delegatum sententia supdca pdcus reverendissimus pater dnus archiepus ac pfati mag' Walterus et frater Thomas pcuratores nõie dctor' dnor' suor' coram eodem dno epo judice delegato judicialit' comparentes unioni, annexioni, sive incorporacioni ecclie, ac concessioni, et assignacioni manerii pdcor' aliisq' omnibus et singlis pdcum dnum delegatum et coram eo ut pmittitur actis habitis atq' gestis consencierunt et consenciunt ac consenciit et consentit quilt eordem, idemq' dnus archiepus p discretum virum dnum Johannem Parker capellanum familiarem ejusdem dni archiepi rectorem ecclie Sti Pancratii London pcuratorem suum ad subsequens pstand' juramentum coram pdco dno delegato in nror' notarior' actor' scribar' et testiũ subscriptor' psencia apud acta hujusmodi constitutum ibm psentem et mag' Walterus et frat' Thomas pcuratores pdci in animas dcor' dnor' suor' ad Sta Dei Evangelia p eosdem pcuratores et eorum quemlt corporalit' tacta corporale pstiterunt juramentum pdcum viz. Johes Parker quod memoratus dnus archiepus et pdci mag' Walterus et fratr' Thomas qd pfati prior aut conventus unioni, annexioni, sive incorporacõi ecclie aut concessioni et assignacõi maner' pdcor' quovismodo contravenire aut eas vel earum aliquam in toto vel in pte infringere non psument aut psumet aliquis eordem; et quia pdcus dnus epus judex delegatus ad ulteriorem executionem in hac pte facienda p tunc ut asseruit intendere non valebat, venerabilibus et discretis viris d̃nis Joħi Elme eccłie pochialis de Lambyth d̃cte Wintonien' et Joħi Parker p̃dc̃o dc̃e eccłie Sc̃i Pancracii Londonien' diocesium rectoribus tunc ib̃m p̃sentibus et cuilibet eorum p se et in solidum ac quibuscunq' capellanis Cantuarien' pvincie conjunctim et divisim ad inducendum p̃fatum d̃num archiep' vel ejus pcuratorem seu pcuratores p eo in corporalem possessionem p̃fati manerii de Woddon juriumq' et ptinenciar' suor' universor' vices suas commisit idem d̃nus delegatus prout harum serie committit cum cujuslibet cohercionis canonice potestate; sup quibus omnibus et singulis p̃fati reverendissimus pater d̃nus archiep̃us et d̃nus ep̃us judex delegatus et p̃dc̃i mag' Walterus ac fratr' Thomas pcuratores requisiverunt nos notarios subscriptos coram memorato d̃no delegato in causa sive negocio p̃dc̃o actorum scribas publicum seu publica instrumentum seu instrumenta conficere p loco et tempore opportuñis. Acta sunt hec prout sup scribuntur et recitantur sub anno indiction' pontificatu mense, die, et loco p̃dc̃is; presentibus venerabilibus et discretis viris mag̃ro Joħe Shillyngford legum d̃re canonico eccłie Wellen', d̃nis Joħe Mandut de Bradestede et Wiłłmo Freman de Plukle Cantuarien' dioc' pochialium eccłiarum rectoribus, et d̃no Wiłłmo Garnonn Lincolnien' et d̃no Wiłłmo Yngylby Roffen' dioc' capellanis, nec non Thoma Burgh et Joħe Grede clericis Eboracen' et Exonien' dioc', Testibus ad p̃missa vocatis specialiter et rogatis.

Licentia regia ad appropriationem et assignacionem eccl'ie de Croydon et maner' de Woddon.

Ricardus Dei gřa rex Anglie et Francie et dominus Hibern'. Omnibus ad quos p̃sentes ĩre p̃venerint saltem. Sciatis q̃d de gřa ñra sp̃ali concessimus et licentiam dedimus ꝑ nobis et heredibus ñris quantum in nobis est dilectis nobis in Xp̃o p̃ori et conventui Sc̃i Salvatoris de Bermondesey q̃d ipsi maneriũ de Woddon cum ptinenciis in com' Surr' q̃d de nobis tenetur in capite ut pcella dotac̃ōis dc̃i p̃oratus de Bermondesey qui de fundac̃ōne ꝑgenitor' ñror' et ñro patronatu existit, dare, concedere, et assignare possint venerabili in Xp̃o patri et carissimo consanguineo ñro Willm̃o archiep̃o Cantuar' in excambiũ ꝑ advocac̃ōne eccl̃ie de Croydon in eodem com' eisdem priori et conventui et successoribus suis p p̃fatum archiep̃um danda et assignanda ĥend' et tenend' dctum maneriũ de Woddon cum ptninent' p̃fato archiep̃o et successoribus suis in excambiũ p̃dct' imppetuũ et eidem archiep̃o q̃d ipse p̃dc̃um maneriũ de Woddon cum ptninen' a p̃fatis priore et conventu in escambiũ p̃dc̃t' recipe possit et tenere p̃dc̃o archiep̃o et successoribus suis imppetuum sicut p̃dc̃um est tenore p̃senciũ similit' licenciam dedimus sp̃alem, Statuto de terris et ten' ad manũ mortuã non ponend' edito seu aliis p̃missis non obstantibus. Nolentes q̃d p̃fati p̃or et conventus vel eor' successores, aut p̃fatus archiep̃us seu successores sui, racione statuti p̃dc̃i seu alior' p̃missor' p nos vel heredes ñros justiciarios, escaetores, vicecomites, aut alios ballivos seu ministros quoscunq' inde occasionentur, molestentur in aliquo, seu graventur, salvis tamen nobis et heredibus ñris serviciis inde debitis et consuetis. In cujus rei testimonium has ĺras ñras fieri fecimus patentes. Teste meipso apud Westmonasterium tercio decimo die Decembr' anno regni ñri quarto decimo.

P bře de privato sigillo,

FARYNGTON.

Hæc indentura facta apud Bermondeseye die Lune in prima septimana quadragesime anno regni regis Ricardi secundi quarto decimo int' venerabilem đnum Wiłłmum archiep' Cantuar' ex una parte & p̃orem & conventum Sc̃i Salvatoris de Bermondeseye ex pte altera, de composicione advocationis vicarii ecclie de Croydon in com' Surr'. Testatur quod collacio & patronatus vicarie đce ecctie de Croydon ad dc̃um đnum archiep̃um & successores suos solum & in solidum in ppetuum ptinebit & spectabit, ad quam quociens cum vacaverit iidem đnus archiep̃us & succ' sui duas ydoneas psonas p̃fatis p̃ori & conventui nominabunt, quar' alteram iidem prior & conventus quam sua discretione duxerint eligendam p̃fato đno archiep̃o & suis successoribus presentabunt ad eandem vicariam p ipsum đnum archiep̃um & successores admittend' & instituend' vicarium, in eadem; ad quã quidem convencionem & composicionem ex pte dc̃i đni Wiłłmi archiep̃i bene & fideliter faciend' p̃dct' đnus archiep̃us obligat se & successores suos & ad quam quidem convencionem & composicionem ex pte dctor' p̃oris & conventus bene & fideliter faciend' p̃dc̃i prior & conventus obligant se & successores suos p p̃sentes. In cujus rei testimonium p̃sentibus indenturis ptes p̃dci sigilla sua alternatim apposuerunt, dicto loco, die, et anno supradict'. Qui quidem vicarius & ipsius singuli successores in eadem ante suas inductiones juramentum manualit' coram priore vel sub-priore & conventu de Bermondeseye in domo sua capitulari p̃stabunt q̃d ipsi ꝑ temporibus suis non impugnabunt p se, alium vel alios clam vel palam arte vel ingenio composicionem seu ordinacionem in aliqua sui pte olim factam p veñlem patrem Johannem nup archiep' Cantuarien' inter rectorem de Croydon & ipsum vicarium & successores suos, vel usurpabunt sibi vel vicarie sue p̃dc̃e aut usurpare pmittent quatenus in eis est aliquod jus vel commodum quod ad p̃dct' religiosos de Bermondsey apud Croydon dinoscitur ptinere, sed erunt eorum adjutores ꝑ posse & eccl̃ie de Bermondeseye fideles.

No. II.

Excerpta ex Computis Ministrorum.

Comp' general' de temp' Edv' II.

Ric' de Fairford, ball' } de Croyndene.
Tho' de Bunchesham, p̃pos' }

In curtilagio fodiendo et plantando .	xiii đ.
In vinea et ł̃bariis reparandis . . .	ix đ.
In iiii surc' ad domos curiæ . .	vi ob̃.
In mcclxxv cendulis clavis, & in conredio, & stip' carpent' coop̃ientiũ warderobam	xv s̃. iii đ.

In bordis & clavis & lattis & in meremio & in stip' carp' repancium salsariam . . . . xxvi đ. ob.
In bordis & lattis, & clavis & stip', & conredio carpentar', cooptor' & recoopientiũ grangias de Croynd' & stablũ & borias & bcaria & pistrin', & in claustro subfulciendo, & in coquina emendanda . . . xxxiiii s̃. viii đ
In emendis xxx carectatis carbonis & cariandis, a Burstowe usq̃ Croydon . liii s̃. ix đ.

Ex comp' Joh' Pieres attornat' Adæ Bochers p̃poi de Croydon, a fõ Mich' 23 Ric. II. ad id' 1 Hen' IV.

In xiim̃[1] de rofnayl empt' p novo stablo & nova cam̃a ibm p m̃[1] xiii đ. . xiii s̃.
Et in cc de sixpenny nayl . . . xii đ.
Et in xxvii carect' zabli empt' ad id' p cujult carect' iii đ. . . . vi s̃. ix đ.
Et in iiim̃l' lyflatch empt' ad id p m̃l vi s̃. xviii s̃.
Et in al' mmm lyflatch empt' p m̃l' vi s̃. viii đ. . . . . xx s̃.
Et in m rechelath empt' . . . v s̃.
Et in xlviii carect' t̃ræ rub' p pariet' dauband' p qualibet carect' . . viii s̃.
Et in i carect' ser̃ais empt' p parietib' daubandis . . . . . xiiii đ.
Et in xlim̃l' tegul' empt' p cooptur' eor̃d' p m̃l' iiii s̃. vi đ. . . . ix t. iiii s̃. vi đ.
Et solut' Robto Kene carpentar', in

partem ɔvenc' suæ fc̃tur' đctor' novor' stab̄lar' & cam' . . . xi s̃.
Et in xxi q̃r' calc' ɔbust' empt' ad id' p q̃rt' xvi đ. . . . . . . xxviii s̃.
Et in cariag' eorđ' p q̃r' ii đ. . . . iii s̃. vi đ.
Et in un' nov' mur' juxt' nov' granar' erga cæmiter' emendand' . . . vi đ.
Itm̃ in iiii^{xx} waynscot bord' empt' p host' & fenestr' nov' stabl' . . . xxiii s̃.
Et ꝑ cariag' eord' de Lond' usque Croydon . . . . ii s̃. iiii đ.
Et in vi regulbord empt' ad idem p pec' vii đ. . . . . iii s̃. vi đ.
Et in md planchisnail empt' ad id p c. vi đ. . . . . vii s̃. vi đ.
Et in m whitnail empt' ad id . . v s̃.
Et in c gross cl' empt' p manger' stabl' xx đ.
Et in l clavis ejusd' sort' empt' ad id' . x đ.
Et in i hõĩe ɔduct' ꝑ i die ꝑ foraminibꝫ in terr' faciend' ꝑ postib' manger' imponend' . . . . . . iiii đ.
Et solut' Rob̄to Kene carpentar', ꝑ ea quæ supius comp' in part' ɔvenc' suæ ꝑ magno stablo . . . . l s̃.
Et in v^{c} whitnail empt' ad id' . . ii s̃. vi đ.
Et in cc gross whitnail empt' ꝑ magno ostio magni stabli et ꝑ magna port' mañii ꝑ c. xviii đ. . . . iii s̃.
Et in rastris cariand' a Bristowe usq̃ Croydon . . . . . vii s̃. vi đ.
Et in i carp' ɔduct' p ii dies ad pen-

dend' mag' portã mañii et pendend' ṽta rastr' in veter' stabl', & etiam emendand' c̃tos defect' palicii circa stagnũ in gardin', p diem vi đ. . xii đ.

Et in ccc saplath empt' ꝑ nov' granar' & ꝑ cam' sup granar' p c. v đ. . xv đ.

Et in i hõĩe ꝯduct' p v dies et dim' p sep' inde faciend' int' angulum coquinæ & stagnũ đni de pco đni p diem iiii đ. . . . . xxii đ.

Et in i cart' ꝯduct' p iii dies tam ꝑmeremio cariand' p le hale ex opposito cellar' vers' ħbariũ quam ex okenstuble, ꝑ nov' pariet' faciend' int' magn' stablũ, et thalam' privat' ad finem stable, p diem xx đ. . . v s̃.

Et in viii nov' estrychbordis empt' ꝑ nov' hostio cellar' ꝑ qualibet' bord' iii đ. ob. . . . . . . ii s̃. iiii đ.

Et in c alb' clavis empt' ꝑ dc̃o hostio . vi đ.

Et in i serrur' cum ob & annul' cum le plate, i latche & i katche empt' ꝑ dc̃o hostio . . . . . . ii s̃. ii đ.

Et in viii q̃r' calc' ust' empt' tam ꝑ mur' dc̃i hostii defect' repand' quam p fundamentis subt' lat' de hale altit' iiii ped' faciend' p q̃rt' xvi đ. . . x s̃. viii đ.

Et in p̃dc̃t' viii q̃rt' cariand' de Halynke & Le Combe ad dc̃t' maner' ꝑ singul' duobus q̃rt iiii đ. . . . . xvi đ.

Et in iii hõĩb' ꝯduct' p i diem, tam ꝑ fundamentis nov' pariet' int' duo stabl' fodiend' quam ꝑ puteis ꝑ post' in terr' fodiend', cuilibet p diem iiii đ. xii đ.

Et in mm teg' plan' empt' tam' ꝑ le hale p̃dc̃t' qua nov' pariet' teguland' p m̃l' iiii s̃. vi đ. . . . . . ix s̃.

Et in mcccccc lath' empt' tam ꝑ đct' nov' hale qua pariet' int' stablũ latand' p c. viii . . . . . x s̃. viii đ.

Et in mmmmmm roffnayl empt' tam p̃đct' hale quam p nov' pariet' in plur' locis dupliciter latand' p m̃l' xiii đ. . vi s̃. vi đ.

Et in c blaknayl empt' ad id' . . . vi đ.

Et in ii q̃rt' calc' ust' empt' ꝑ cooptura de la hale & đct' nov' pariet' p q̃rt' xvi ii s̃. viii đ.

Et in cariag' ejusdem . . . . iiii đ.

Et in ii q̃rt' zabli fodiend' & cariand' ad id' . . . . xvi đ.

Et in i tegulat' cũ garcione suo ꝯduct' p x dies p le hale & đct' pariet' teguland' in gross viii s̃. iiii đ. int' se p diem x đ. . . . . . . viii s̃. iiii đ.

Et in i carpentar' ꝯduct' p i diem p rackes in vet' stabl' emendand' . vi đ.

Et in vi carect' t̃ræ rub' cũ cariag' empt' ꝑ đct' pariet' dauband' tam de la hale quam nov' pariet' p̃đct' ꝑ qualibet carect' ii đ. . . . . . . xii đ.

Et in ii hõĩbus ꝯduct' p iiii dies ꝑ đct' hale & pariet' dauband' cuilib' p diem iiii đ. . . . . ii s̃. viii đ.

Et in stip' Rob' Kene carpentar', ꝑ đct' nov' pariet' faciend' ex ꝯvent' fc̃ta in gross . . . . . xx s̃.

Et solut' eid' Robto de denar' sibi deb' de anno pced' p fcura nov' stabli ex 9ven' fca in gross . . . xx s.

Et solut' eid' p dict' hale faciend' ex 9ven' fca in gross xxx s. unde p senescall' hospitii xv s. . . . xxx s.

Et solut' Wiltmo Mason p fundamentis de dct' hale faciend' ad altit' iiii ped' ex 9vent' fact' unde p senescall' hospit' vi s. viii d. . . . . . xiii s. iii d.

Et in xxvi teg' concav' empt' p pariet' pdct' crestand' . . . . . xiii d. ob.

Et in i bushell tigulpyn emp' . . vi d.

Et solut' Wiltmo Mason p i nov' hostio petrar' de caine faciend' & ponend' ex 9vent' fact' in gross xii s. unde p senescall' hosp' vi s. viii d. . . xii s.

Summa xxxiii lib' xiiii d. und' p senescall' hosp' . . . xxviii s. iiiid.

An Imperfect Roll, de ann. 34 Hen. VI.

In divsis expens' hoc anno fcis sup repacoe m. de Croydon, ut in denar' solut' p cariag' xxi ml' tegul' de Bewle usq, man' pdcum . . xiiii s.

c rofetyle . . . . . . iii s. iiii d.

x bz tylepynnes p bz xv d. . . iiii s. ii d.

c hertlath . . . . . . vii d.

Clav' de divsis sortibus . . . xxi s. x d.

xxx qrt' calc' ust' p qrt' xiiii d. . . xxxv s.

Simul cũ fodic͠ōe zabul' & car' ejusd' . iiii s̃.
iiii l. ii s̃. xi đ.

Et in denar' solut' Robto Tyler & iii soc' suis, laborant' sup repac͠ōe domor' infra m̃ ibm, viz. sup cooptur' eãdem nec non iiii° laborariis eisd' tegulatoribus servient' in opibus predc͠is dĩvs' vic' p̃dict' ꝯduct' infra' dc͠m temp' hujus comp̃i . . . iiii l. xvii s̃. ob.

Et in denar' solut' Johi Wylde tegulatori & servient' suo, opant' sup rep̃ac͠ōẽ domor' infr' m̃ p̃dc͠um p iiii đ' dies ad xiiii đ. p diem inter se . iiii s̃. viii đ.

Et in denar' solut' Adæ Pykman, p ꝯduc͠ōe caractæ suæ cũ ii laborar' laborant' in cariand' unã bigat' lapid' voc' fryseton de Mestħm usq' m̃ p̃dcum . . . . . iii s̃. iiii đ.

Et in denar' solut' Tho' Wareham carpent', p dĩvs' laborib' p ip̃m fc͠is infra m̃ p̃dc͠m, viz. p fc͠ura opis carpentriæ lect' đni cũ merem' & tabul' empt' ad id opus, nec non p repac͠ōe dom' carbonũ simul cnm emendac͠ōe dĩvs' defectuũ aliar' domor' ibm & fc͠tura dĩvsor' necior' hoc anno . xxvii s̃. ii đ.

Et in denar' solut' eid' Tho' Wareham, p mmdcccc hertlath, p c. vii đ. xvi s̃. xi đ. et cccl ped' de evesbord p c. xx đ. v s̃. x đ. empt' et expendit' sup rep̃ac͠ōẽ domor' m̃ ibm . . xxii s̃. ix đ.

Summ' xi l. xvii s̃. x đ. ob.

Et in divs' exp' hoc ann' fcis sup repacōe dom' de le portmote situat' infr' vill' de Croydon, ut in meremio, tegul', calce, zabulo, clav' de divs' sortib', luto, & aliis rebus empt' & expendit' in op' pdco, simul cū ꝯduccōe carpentar', tegulator', daubator' & alior' laborar' ibm laborant' p divis' dies infra temp' hujus compi . xl s.

Et in denar' solut' p conduccōe unius carectæ cū duobus laborar' laborant' in cariando palic' de bosc' apud Waldyngham usq' clausur' parci de Croydon p ii dies ad xx d. . . . . iii s. iiii d.

Computus Adæ Pykman & Ric' Pykman pcarum de Croydon. Without date, but probably subsequent to 33 Hen. VI.

Et in denar' solut' p refect' xlix pticat' vet' sep' clausur' pci pdci in divs' locis pci ibid' hoc anno, unde ex opposito le Quas hoc xxxix ptic', ex opposit' le Brake v ptic', & versus le Pondes v ptic', singul' ptic' i den' . . . . . . iiii s. i d.

Et solut' Ricardo Kyppyn facienti et opanti xxiii ptic' et dim' nov' sepis juxta le Pound. singul' pticat' ii d. ob. . . . iiii s. viii d.

Et solut' eid' Ricardo p fctur' xxviii pticat' nov' sepis fct' apud le Bromhill p pticat' ii d. ob. . . . . v s. x d.

Et solut' eid' Ricardo p fcur' vi pticat' nov' sepis fact' apud Horsepondfold . xv d.

Et eid' ꝑ fctura ii pticat' nov' sepis erga le Gretbrake . . . . . . v đ.

Et solut' eid' emendant' et reficient' xlvi pticas vet' sepis p ipsū fce apud Toderys crod' pticat' ad i đ. . . iii s̃. x đ.

Et solut' eid' Ricardo ꝯduct' p iiii dies sup emendacōē defect' in dīv̄s' locis pci p̃dc̄i p diem iiii đ. ob. . . . . xviii đ.

Et solut' Wiltmo Atte Hethe & Nicolao Cooper, p lxxxviii ptic' nov' palic' p ipsos fc̄t' in le Rowting ptic' iiii đ. ob. . . xxxiii đ.

Et solut' p ꝯduccōe x bigat' car' đct' palic' a Tillingdowne usq' pcū đni p i diem cuilt eor' p diem ii s̃. . . . . xx s̃.

Et solut' ꝑ car' vi bigat' des postes, railes, & shorys de okestob' ad pcū đni qualt' bigat' vi đ. . . . . iii s̃.

Et solut' ꝑ fc̄ur' duor' novor' psepiū ordinat' ꝑ feno imponendo ad dam' pascend' in hiem . . . . . . iii s̃. vi đ.

Et solut' ꝑ cariagio đcor' psepiur' a Sandteswad usq' in pcū p̃dc̄ū . . . vi đ.

Et solut' ꝑ cariagio unius bigat' feni a prato usq' pcum p̃dc̄um . . . . viii đ.

Et solut' i laboratori ꝯduct' cū i equo tractante spinas & subboscum ad manus operarii p vi dies p diem v đ. . . ii s̃. vi đ.

Summa iiii t. v s̃. ix đ.

Comput' Ric' Pykman, Cust' m̃ de Croydon a fo' Mich' 6 Ed. IV., ad id' 7 Ed. IV. Ex Comp' General' istius anni.

In divs' expens' hoc anno fcis sup repacõe domor' m̃ ibm ut in denar' solut' p iiii^{or} m^{l} d tegul' . . . . xviii s̃.
v q̃rt' iiii bz calcis ustæ p q̃rt' xiiii đ. . . vi s̃. v đ.
Stipend' unius tegulatoris p xviii dies . . ix s̃.
Et servientis sui p idem temp' . . vi s̃.
Alterius tegulatoris p xxiiii dies ad v đ. p diem. . . . . x s̃.
xxiiii l. soldure p̃t' lb. iiii ob. . . . ix s̃.
Una bigat' arenæ . . . . iiii đ.
ii big' zabuli . . . . . vi đ.
Una big' luti . . . . . iii đ.
Stipend' unius daubatoris p iiii dies . . xviii đ.
c findul' . . . . viii đ.
m rofentyle . . . . . . viii đ.
Empt' et expend' sup reparacõe domor' m̃ p̃dci, lxii s̃. ii đ.
Et sol' p reparacõe sup dom' portmote ibm . (*torn*)

Computus Joh' Lytyll Cust' m̃ & pci de Croydon, a fo' Mich' 13 Ed. IV. ad id. Ex Comp' Gen'al' istius anni.

In expens' hoc ann' fcis sup repacõẽ m̃ ibm ut in denar' solut' p v q̃rt' calcis ustæ . v s̃. x đ.
mmm tegul' cũ cariagio . . . . xii s̃.
Factura de les rakkes & mangers in stabulis ibm . . . . iii s̃.
v bigat' zabuli . . . . . . x đ.

ii bigat' luti . . . . . . vi đ.
Stipend' unius daubatoris p ii dies . . x đ.
Erectio unius posti in gardino . . ii đ.
Et tegulatio muri in gardino ibm . . vi đ.
Pro clavis de divsis sortibus . . . v đ. ob.
Summa xxiv s. i đ. ob.

Et p iiii novis clavis p divs' hostiis ibm, & emendacōe ii serrar' ibm . . . . . .
Summa xxv s. ix đ. ob.

In divs' expens' fcis sup clausur' pci ibm, ut in denar' solut' p fcura c paxillar' . . iiii đ.
Carig' eard' de Okestubble . . . v đ.
Circa i bigat' de Edders . . . v đ.
Tractur' de Tynet . . . . . ii s. vi đ.
fcura xlii ptic' sepis ibm . . . x s. vi đ.
Et fcura xxiiii ptic' sepis circa prat' đni . ii s.
xvi s. ii đ.

A Generall Roll. Imperfect, without date, circa 14 *Ed. IV.*

Cust' M. de Croydon.

In denar' solut' Tho. Warham carpentar', p fcura opis carpentriæ stall in foro de Croydon de novo ex ɔvencōe secū fca in grosso . . . . liii s. iiii đ.
Et sol' Joħi Fermour Shyngler p po sicōe vi m̃. shyngle sup đct' stall p cooptura eor' p mill' iii s. viii đ. . . . . xxii s.
Et sol' pfato Tho' Warham p fctur' de lez

bynnes in panetria ꝑ pane impanend', ex ꝯvencōe in grosso . . xxxiii s̃. viii đ.

Et sol' eid' Tho' ꝑ fc̃ura & posicōe unius somer in le pastry ib̃m, & fc̃ura unius muri desup . . . . x s̃.

Fc̃ura unius copborde in aula . iii s̃. iv đ.

Et ꝑ diṽs' opibus fc̃is sup altare in capella ad ponend' jocalia desup . ii s̃.

Summa xv s̃. iiii đ.

Et sol' eid' ꝑ meremio cum sarratione ejusd', & ꝑ findul' ꝑ p̃stall' in foro v s̃.

Et sol' Joh̃i Plomer ꝑ soldur' & emendacōe gutter in diṽs' locis m̃ . . viii s̃.

Et in denar' solut' Rob' Tyler opant' sup cooptur' domor' m̃ ib̃m . . xiii s̃. iiii đ.

Et sol' ꝑ ii clavibus ꝑ hostio columbar' ib̃m . . . . iii đ.

Comp' Joh. Lyttyll Cust' m̃ pci de Croydon a fo' Mich' 22 Ed. IV., ad id' 1 Ric. III. Ex Comp' Geñal' hujus anni.

In expens' hoc anno fc̃is sup rep̃aracōẽ domor' m̃ p̃dc̃i prout patet, &c. . . lx s̃. iii đ.

Et in denar' solut' ꝑ fc̃ura unius stadii palic' p clausur' pci ib̃m hoc anno . . . x s̃.

Summa lxx s̃. iii đ.

General Roll. Imperfect, without date, temp. Hen. VIII.

In denar' solut' ꝑ divs' repacōib' hoc anno fact' sup man' de Croydon, viz. ꝑ fcura iiii^or portar' cū suis ptinen', ac ꝑ vadiis div' sor' carpentarior' tegulator', & alior' laborator' cum divsis necessariis . . . . . . x ł. xv s. ii đ.

Et in denar' solut' ꝑ fcura ii furlong paliciæ circ' pcū ibm, cum cariag' ejusd' palic' . . . . xxix s. ii đ.

Comp' Christoph' Hore gen', præpos' m̄ de Croydon a fo' Mich' 1644, *ad fm̄ Mich'* 1645. *Ex Rot' General. istius anni.*

Et de 40 ł. de firma situs palatii cū omnib' dom', ædif' stabul', columbar', gardin', pomar', piscar', & al' ptin', nup in manu ꝑpia archiep' Cant' et modo comitis Nottingham.

Et de 66 ł. 13 s. 4 đ. de firma pci cū cappic', bosc', subbosc', & terr' boscal', vocat' le Parke coppice, cont' insimul p estim' 170 acras, una cū dom' mansional' in eadem, ac horr', stabul', & al' edific' ad eandem spectantib'.

No. III.

Extract from the Minister's or Bailiff's Accounts of the Colleges, Chantries, Free Chapels, &c., in the County of Surrey, 3 Edw. VI., deposited in the Augmentation Office at Westminster.

Nup' Canteria b'te Marie in Croydon. } Et r' compm de xx s̃. de firma unius teñti ibm juxta le Church-gate cū ptinen' in tenura Thome Comporte, p ann' ut supra sol' ad ii anni t̃mos, viz. a festo Annunc.' b̃te Marie virḡis et Sc̃i Mich̃is arch̃i; de vi s̃. viii đ. nup rec' de firm' domus mansionis Cant̃ie p̃dct' cū om̃ibz et sinḡlis s' ptin' modo vel nup in tenura incūbent̃z ibm videlt p tempus hujus comp̃i non r', eo q̃d conceditur Willmo Warde & hered' suis impp̃m a Festo Sc̃i Mich̃is arch̃i anno regni đni regis nunc Edwardi VI^{ti} sc̃do, put p l̃ras ejusdem dñi regis paten' dat' xviii° die Aprilis anno regni mat̃s sue iii°, tenend' de p̃fato đno rege, hered' et success' suis in libo socagio absq; x^{ma}, ut in eisdem l̃ris patent' plenius pz. Sed r' de vi s̃. de firm' unius teñti jacen' in Pickelake cū ptin' in tenura . . . . Robsonne p ann' sol' ut sup̃. Et de xvi s̃. de firm' unius teñti cū ptin' in tenura Joh̃nis Curts, p ann' sol' ut supra. Et de viii s̃. de firm' unius teñti in tenura Ric̃i Alford p ann' sol' ut sup̃. Et de xx s̃. de firm' iii cotag' cū ptin' in tenur' Thome Thornetonne p ann' sol' ut sup̃. Et de xii đ. de firm' unius gardini in bor' pte ville de Croydon jacen' juxta teñtum Rici Draps p ann' sol' ut supra. Et de vi s̃. viii đ. de firm' unius teñti ibm in tenura Thome Parker p ann' sol' ut supra. Et de ii s̃. de firm' unius cotag' jacen' int' teñtum voc' Le Crowne p ann' sol' ut supra. Et de iiii s̃. de firm' unius teñti jacen' int' shopam ptin' ad templum de Croydon in le Bocherrowe p ann' sol' ut supra. Et de xv s̃. de firm' iii teñtor' in le Bocherrowe in tenura sol ut supra. Et de xv s̃. de firma unius teñti in foro in tenura Francis Reswid p ann' sol' ut supra. Et de xiii s̃. iiii đ. de firma unius teñti in tenura Johis Baldwyn. De xix s̃. viii đ. nup rec' de reddit' divs' pcell' terr' et teñt' subsequen', videlt de firma unius croft ibm cont' v acr' juxta pcum de Croydon in tenura Johis Hatcher ad x s̃. p ann'; firma unius horrei ibm in tenura Roberti Comporte vi s̃. viii đ. p ann'; firma unius gardini apud Stakecrosse in tenura Johis Reade ad ii s̃. p annum; et de firm' unius acr' jacen' apud Addescombe modo in occupacõne Elizabethe Herne vid' ad xii đ. p ann' scil' p tempus hujus compi non r'. Eo q̃d omnia et singla p̃miss' cũ ptin' int' al' concedunt' Thome Reve et Georgio Colton ac hered' eor' imppm, a festo Sti Michis archĩ anno regni dni regis nunc Edwardi VIti scdo, put p p̃z lras ejusdem dni regis paten' dat' xmo die Maii anno regni mais sue iiicio, tenend' de p̃fato dno rege & success' suis in libo socagio ab qxms, ut in eisdem lris paten' plenius patet sed r' de xii đ. de firma unius acr' terr' jacen' in quodam campo vocat' Teyntefeld in tenura p̃dci Johis Hatcher. Et de vii s̃. de firm' iiii acr' di' terr' in campo voc' Brechefeld in ten' p̃dct' Johnnis Hatcher. Et de vi đ. de firm' di' acr' terr' jacen' in campo voc' Teyntefeld p̃d' in tenura supdc̃i Joħis Hatcher. Et de xii đ. de firma unius gardini jacen' in le olde towne in tenura .... Et de xiii s̃. iiii đ. de firm' iii acr' terr' insimul jacen' in Waddon Warshe in tenura Roberti Crostinge p ann'. Et de ii s̃. de firma unius acr' ter' jacen' in Northstakefeld in tenura Wiłłmi Tomson p ann'. De vii ł. nup rec' de reddit' dĩvs' teñtor' subsequen', videłt de firma ii teñtor' in London in pochia Sc̃i Micħis in Cornehyll in tenura ...... Dommer ad lxvi s̃. viii đ. p ann' firma; unius teñti in pochia in tenura relicte Nicħi Wedouz ad xlvi s̃. viii đ. p ann'; et de firma unius teñti in Trynytie Lane in London, in tenur' Jacobi Chastleyn ad xxvi s̃. viii đ. p ann', sciłt p tempus hujus comp̃i non r'. Eo q̃d omnia p̃missa cũ eor' ptin' int' al' concedunt' Thome Watson et Wiłłmo Adys ac hered' eor' impp̃m, a Festo Sc̃i Micħis Arcħi anno regni đni regis nunc Edwardi VI^{ti} sc̃do, put p p̃z łras ejusdem regis paten' dat' xxv die Marcii anno regni mat' s' iii^{cio}, tenend' de p̃fato đno rege, hered' et succ', in lib̃o burgag' civĩs ñre Lond' absq, x^{ma}, ut in eisd' łris pat' plenius patet.

Sm^{a} vii ł. xv s̃. x đ.

Nup' Cant'ia S'ci Nich'i in Croydon. Nec r' xxxvii s̃. ii đ. nup rec' de reddit' dĩvs' pcellar' terr' & teñt' subseq', videłt de firma domus mancionis Cant' p̃d' cũ ptin' modo vel nup in tenura incumbeñs ib̃m ad vi s̃. viii đ. p ann' firm' di' acr' pastur' in Benshamfeld ad vi s̃. p ann'. Et de firma unius messuagii cũ ptin' in tenura Thome More ad xxx s̃. p ann' sciłt p tempus hujus comp̃i non r'. Eo q̃d omnia et singla p̃miss' cũ eor' ptin' int' al' conceduntur Anthonio Aucher militi et Henrico Polsted ac hered' eor' imppm, a festo Sci Michis archi anno regni dni regis nunc Edwardi VI[ti] secundo, et non ulterius, put p pz tras ejusdem dni regis patent' dat' x° die Marcii anno regni mats s' iii°, tenend' de pfato dno rege, hered' et successor' s', in libo socagio absq, x[ma], ut in eisdem tris paten' plenius patet'. Sed r' de iiii l. de firma unius hospicii voc' Le Crowne cu iiii acr' terr' ac' ix swathes pdci & unius gardini itm p ann'. Et de xxxiii s. iiii d. de firm' unius mess' cu ptin' in Chelmerden de iiii l. ii s. nup annuatim rec' de redd' et firm' divs' mess' terr' et tent' subscript', videlt unius mess' cu ptin' in tenura Johis Pratt ad xx s. viii d. p ann', altius tenti cu vii acr' terr' in tenura Robti Ingram ad xl s. p ann', unius tenti cu ptin' in tenura Johis Fisher ad xiiii s. p ann' ult' ii s. p uno pvo clo non vendit' videlt p tempus pdcum non r'. Eo qd dnus rex nunc Edwardus VI. p tras s' paten' dat' xviii die Aprilis anno regni s' mats iii[mo], int' al' dedit & concessit eadem pmiss' cu ptin' Wiltmo Warde & hered' s', a festo Sci Michis archi anno ii° domini regis pdci imppm absq, x[m] sive aliquo alio pinde reddend' solvend' vel faciend' put in eisdem tris paten' plenius pz'. Sed r' de x s. de firm' unius mess' cu gardino in tena Edwardi Cowper, de xii s. nup ret', de firm' unius tenti cu ptin' in tena Wiltmi Milles geno, scilt p tempus hujus compi non r'. Eo qd dnus rex nunc Edwardus VI[tus], p tras s' paten' dat' x° die Maii anno regni s' mats iii°, int' al', dedit & concessit dcum mess' cum omibz s' ptin' Thome Reve et Georgio Cotton & hered' s' a festo Sci Michis archi anno scdo dci dni regis impp' absq, x[ma] sive aliquo alio pinde reddend', solvend', vel faciend', put in eisdem tris paten' plenius patet. Sed r' de ii s̃. p redd' unius pvi clausi in teña Joħis Fyssher ut p̃z dc̃m librum. Et de xx s̃. de firm' unius teñti ib̃m voc' le Brodgate cum ptin' in tenura Joħnn' Lane. Et de x s̃. de firm' unius teñti cum ptin' in tenura Johann' Crowne. Et de vi s̃. de firm' unius shap in le Fishe m̃kett in tenura Roberti Wrythesley, ut p̃z dc̃um librum supvis'.

S'ma viii ł. xvi đ.

No. IV.

[*From* "Chartæ Miscellaneæ," *in the Archives of Lambeth, Vol. XIII. No.* 16.]

Com' Sur'.

Parcell of the Possessions of the late dissolved Archbishopricke of Canterbury.

All those free rents of assize belonging to the said mannor of Croyden, per ann.	lviii ł. xiii s̃. vi đ.
All those ffines of boroughholders .	xxiiii ł. xi. ob̃.
The farm of all that the scyte mannor or . . . . se of Croydon, with a chappell wainscotted, a grainary, and all houses, outhouses, court yards, and other yards and stables thereunto belonging, encompassed	

with a faire court yard on the north, a small running water east and south, and the church yard west, and of all that great garden and fruite howse, with all other gardens, orchards, pidgeon howse, waters, and three fish ponds fenced on the west with a brick wall on the south, with a small running water and a hedge on the east, with another hedge belonging to the howse keepers meadowe, on the north, with a water which parteth the aforesaid gardens and the said meadows, per ann. . . . . xl li.

The farme of all that meadowe called Birch Meadowe, contayninge by estimason 4 acres and an half, per ann. . . . . . . vi li. xv s.

The farme of all that meadowe called Stubbs Meadow, containinge, by estimason, 30 acres, per ann. . xxx li.

The farme of all those two meadowes called Northbury Meadows, containinge 18 acres, per ann. . xvi li. xiii s.

The farme of all that howse called the Parke Howse, situate and being in Croyden Parke, with all howses, one barne, one stable, one orchard and garden thereunto belonging, containinge in the whole two acres or thereabouts, per ann. vi li.

The farme of all that feild called Hillyfield, lying on the south of the said Park House, containinge, by estimation, 6 acres or thereabouts, per ann. . . . . . . xxx s̃.

The farme of all that meadowe called Parke Meade, containinge 16 acres, per ann. . . . . . . xi li. iiii s̃.

The farme of all that close of pasture called Mareclose, containinge 11 acres, per ann. . . . . . cx s̃.

The farme of all that feild of pasture called Oakefeild, containinge 6 acres, per ann. . . . . . lx s̃.

The farme of all that feild of pasture called Roundhill, containinge 11 acres or thereabouts, per ann. . lxxiii s̃. iiii d.

The farme of all that feild of pasture called Wheatefeild, containinge 4 acres and an half, per ann. . xlv s̃.

The farme of all that feild of pasture called Ruttingefeild, containinge 18 acres, per ann. . . . . ix li.

The farme of all that feild of pasture called Layfeild, containinge 8 acres, per ann. . . . . . . iiii li.

The farme of all that feild of pasture called the Lowefeild, containinge 4 acres, per ann. . . . . . xl s̃.

The farme of all that field of pasture called Lower Goers, containinge 10 acres, per ann. . . . . . c s̃.

The farme of all that field of pasture called Cowpast, containinge 13 acres, per ann. . . . . . vi l. x s̃.

The improvement of Bushy Parke, being grubbed, per ann. . . . x l. xv s̃.

The farme of all that field called Bushy Parke, containinge 34 acres, 3 roods, and 20 perches, per ann. . vi l. v s̃.

The farme of all that close called Oatefield al' Pickedfeild, containinge 6 acres, 1 rood, 24 perches, per ann. . . . . . . lx s̃.

The farme of the vesture of all that coppice called the Parke Coppice, containinge 15 acres, per ann. . lxxv s̃.

The farme of the vesture of all that wood called Biggin Great Coppice, containinge 64 acres, 3 roods, 21 perches and an half, per ann. . ix l. xv s̃.

The farme of the vesture of all that coppice called Bewdly, containinge 22 acres and 20 perches, per ann. lxxiii s̃. iiii d.

The farme of the vesture of all that coppice called Windalls, containinge 21 acres and 30 perches, per ann. . . . . lxx s̃.

The farme of the vesture of all that coppice called Shelverdins, con-

taininge 18 acres, 1 rood, and 10 perches, per ann. . . .	lx ł. xi š.
The farme of the vesture of all that coppice called Stakepitt, containinge 71 acres and 30 perches, per ann.	xi ł. xiii š. i đ.
The farme of the vesture of all that coppice called Gravelly Hill, containinge 27 acres, 2 roods, and 27 perches, per ann. . . .	iiii ł. xi š. viii đ.
The farme of the vesture of all that coppice called Little Stakepitt, containinge 8 acres, 2 roods, and 33 perches, per ann. . . .	xxix š.
The farme of all that lodge in Croydon Parke, with a barne, stable, outhouses, garden, and orchard thereunto belonging, after the death of Francis Lee, Gentleman, who now holdeth the same during his life, by a patent thereof, dated the 25th day of November, a° 1637, as keeper of the said parke and reeve of the woods belonging to the said archbishopricke of Canterbury, in the county of Surrey, per ann. .	Nil.
The profits of two faires to be kept .	xii š
All those the royalties of hawkinge, huntinge, fowlinge, fishinge, courts leet, courts baron, liberties, fines, issues, amerciaments, heriots, waifs, estrayes, escheats, deodands, tre-	

sures trove, felons' goods, and wrecks of sea, &c. . . . Nil.

The perquisites and profits of courts leet and courts baron belonging to the said manor . . . . Nil.

The mannor of Croyden is valued to be worth cclxxiiii li. xix s. ix d. ob.

Timber and Underwoods.

Fifty-two elms, 5 birch trees, and 1 ash, neere about the mansion-house of Croyden aforesaid . . . . . xxx li.

Timber growinge and being upon the aforesaid field called Cowpasture . cc li.

Timber pollards and bushes growing and being upon Bushy Parke aforesaid . cccc li.

Timber growinge and beinge upon the parke coppice . . . . . . xl li.

Woods and underwoods groweinge and being upon the parke coppice aforesaid xxx li.

Woods and underwoods groweing and being upon Biggin great coppice aforesaid . . . . lxxi li.

Woods and underwoods groweing and being upon Bewdley coppice aforesaid xxii li.

Timber and pollards there . . vii li. x s.

Woods and underwoods groweing and being upon Windalle's coppice aforesaid . . . . xxi li. v s.

Timber and pollards there . . x li.

Woods and underwoods groweing and being upon Shelverden's coppice aforesaid	xxi li. xi s.
Timber and pollards there . . .	vi li.
Woods and underwoods groweing and being upon Great Stakepitt aforesaid .	xxxv li. xii s.
One hundred and fifty small trees, or thereabouts, being timber and pollards growinge and being upon Great Stakepitt aforesaid . . .	xxx li.
Woods and underwoods growing and being upon Gravelly Hill coppice aforesaid	xliiii li.
Thirty small timber trees and pollards there	lxxv li.
Woods and underwoods growing and being on Little Stakepitt aforesaid .	xxix s.
Timber and pollards growing and being there	iiii li. x s.
Eighty timber trees growing and being upon one wood called Norwood, lying in the parish of Croyden, and part thereof in the parish of Lambeth .	l li.
Nine thousand two hundred oaken pollards, with the tops and lops, growing and being upon the wood aforesaid, which wood containeth eight hundred and thirty trees	mmccxl li.

The small spray of all pollards, when the same are lopped for coaling, the small spray of every coppice wood, as they are felled, and all dotard and rotten trees, and the bark of all trees yearly felled, being an increase of profit to be made of the woods after the death of the said Francis Lee, which is certified to be granted to him during his life, by the patent above mentioned . Nil.

Timber and timber-like trees upon the lease lands of Adam Torlis, Gentleman . . . . iiiiviii l. xiii s iiii d.

Reprizes uncertain.

To Francis Lee, by patent dated the 25th November, 1637, as keeper of the park of Croyden, and reeve of the woods in the county of Surrey, all that lodge in the park of Croyden, with a barn, stable, outhouses, garden, and orchard thereunto belonging, for his life . . . . Nil.

To him, for grass for two cows or oxen in the park, per ann. . . . . x l.

To him, for a fee of two pence per diem, per ann. . . . . . . lx s. x d.

To him, more, the small spray of all pollards, when they are lopped for coaling, with the small spray of every coppice wood as they are felled, with dotard or rotten trees, and the bark of all trees yearly felled, and allowance for making and mending the fences about the woods which is not valued . Nil.

To Ralph Watts, Gentleman, and Sir

George Askew, Knt., all profits, trees, and advantages belonging to the housekeeper's place, granted to them by patent, dated the 10th November, 1630. Nil.

To Walter Dobson, Gentleman, for his life, as steward of the archbishop's liberties in the county of Surrey, the moyety of all casuall profits of the said archbishop's royaltyes within the said county, granted to him by patent, dated the 25th November, 1632, and by him assigned on the 3rd day of June, 1636, to Mr. Watts . . . .

To John Dendy and Thomas Smith, during their lives, as bailiffs of the mannor and libertyes of Croyden, a fee of iiii ł. yearly, granted by patent, dated the 25th January, 1638, to be issuing out of the profits accrewing to the said mannor, and paid at Lady-day and Michaelmas half-yearly, and liberty to enter and distrain upon any part of the said manor, for default of payment thereof . . . . . . iiii ł.

xvij ł. x s̃.

Mem.—The premises (except the Gatehouse) were in the possession of the late Archbishop of Canterbury, and are parcel of the possessions of the said late bishoprick

The extents of the manors are not certified.

The quitt rents are collected by reeves yearly chosen of the tennants, and payd in without allowance.

Upon the death of every freeholder, one yeare's quitt rent is to be payd.

Upon alienason, one penny is due by the purchaser for fealty, but nothing for releif.

Upon the death of every coppyholder dying seized of any coppyhold lyinge without the four crosses, there is due the best beast for a heriott; if he have none, then 3 s̃. 6 d̃. in lieu of a dead heriott.

If a stranger purchase any of the coppyholds, the fine is uncertaine; but, upon a descent, the heir is to pay one year's quit rent for a fine.

Upon every surrender to any use or uses, 3 s̃. 6 d̃. is to be paid in lieu of a heriott.

The several uncertaine profits before mentioned are not valued *communibus annis* by the survey, nor any other perquisites or profits of courts or royalties, but it is certified that several beadles are to be chosen out of the tenants, who are to collect and accompt for the same, without allowance.

The Lord's interest in Shurley Heath common, containing 300 acres, and in Croyden Heath, containing 340 acres, and in Thornton Heath and Broad Green, containing 20 acres, (in all which commons the tenants have common sans stint), is not valewed by the survey.

The Lord's interest in Norwood, containing 830 acres, wherein the inhabitants of Croyden have herbage for all manner of cattle, mastage for their swine without stint, and all furze, bushes, broome, and underwoods, is not valewed by the survey.

The tenn pounds fifteen shillings improvement for Bushy Park is over and above the rent reserved; but without the same be first grubbed, the said improvement is not to be made.

The yearly value of the premises, with the improvement of Bushy Park, per ann. 274*l.* 19*s.* 9½*d.*

The value of timber pollards and underwoods is 3356*l.* 15*s.* 3*d.*

The total value of the reprises, during the lives of Francis Lee, John Dendy, and Thomas Smith, is 17*l.*

The materials of the house to be taken down and sold, are valewed at 1200*l.*

It is not certified that these values are as the same were worth in the year 1641.

All the advowsons, and all rights of presentation to any church or chapell, and all donatives, and also all rent and rents reserved upon any demise in being of any lands or tenements, part or parcell of the said manor, together with the said manor and tenements not specified in this particular, and the value of the demesnes, and all rents in arrear at Lady-day now last past, for or by reason of the premisses, and likewise all charters, deeds, evidences, or writings, any ways touching or concerning the same, are to be excepted.

This particular is grounded on a survey taken 17 March, 1646, by Edward Boyer, Esq., and others, and is made forth by order of the contractors of the 12th of April, 1647.

No. V.

[The books from which the following extracts are made are deposited in the Lambeth Library.]

In a Book marked at the back " Au. D. N° 4. II." p. 339.

Croydon, Jan. 26, 1652.

In pursuance of two several orders of the committee for reformation of the universities, of the 15 Jan. 1650, and 28 Jan. 1651, as also an especial order of the said committee, it is ordered that Mr. Lawrens Steele, treasurer, doe from time to time continue and pay to Sir William Brereton the sum of 50*l.* for the use of such ministers as have been and shall be by him provided to serve the cure of the church of Croydon in the county of Surrey, the same to be continued till further orders of the said trustees, and to be accompted from the last receipt, any order to the contrary notwithstanding.

John Thorowgood, William Skynner, William Steele, Richard Younge, John Pocock.

Ibid. p. 340.

Croydon, March 4, 1653.

Ordered, that Mr. Lawrens Steele, treasurer, do continue to pay unto Sir William Brereton the yerely sum of 50*l.*, for the use of such as have been and shall be by him provided to serve the cure of the church of Croydon in the county of Surrey, according to the orders of the said trustees, of the 26th January last.

John Thorowgood, William Steele, Richard Younge, Nic. Martin, John Pocock.

In a Book marked "Au. G. N° 7, 8," fol. 689.

Croyden, May 31, 1654.

In pursuance of two several orders of the committee for reformation of the universities, of the 15 Jan. 1650 and 28 Jan. 1651, it is ordered that Mr. Lawr. Steele, treasurer, doe from time to time pay unto Sir William Brereton, for the use of Mr. Jonathan Westwood, minister of Croydon, in the county of Surrey, approved according to the ordinance for approbation of publique preachers, the yerely sum of 50*l.*, for and during such time as the said Mr. Westwood shall continue to discharge the duty of the minister of the said place, till further orders of the said trustees, together with all arrears of the said 50*l.* per ann., due by order of the 26 Sept. 1652.

John Thorowgood, William Stele, John Browne, Richard Yong, John Powick.

In a Book marked "Au. I. 13. 7," *fol.* 171.

I find 25*l.* paid to Jonathan Westwood, minister of Croyden, charged upon the revenues of South Bersteed, in Sussex, by an order dated April 24, 1655.

Signed by John Thorowgood, Edward Cresset, Richard Sydenham, Edward Hopkins, Ra. Hall, Richard Yong.

Ibid. fol. 226, *Aug.* 13, 1655.

An order to Mr. Lawrence Steele, treasurer, to pay unto Mr. Jonathan Westwood, minister of Croydon,

30*l.*, towards satisfaction of the arrears of the augmentation to him due from these trustees; and Mr. John Silverwood, receiver, is forthwith to pay to Mr. Lawrence Steele, treasurer, 30*l.* for that purpose.

John Thorowgood, Edward Cresset, John Pococke, John Humfrey, Richard Yong.

Ibid. Oct. 11, 1655.

A further order, dated Oct. 11, 1655, to Mr. Allen, receiver, to pay to Jonathan Westwood 4*l.* 8*s.*, charged upon Bartholomews without Chichester, in full satisfaction of all such augmentations, allowances, and payments, as he standeth charged to pay unto them since 25 March last. The payments to be made unto them to be accounted from their last receipt only.

John Thorowgood, Ralph Hall, R. Sydenham, John Humfrey, John Pococke,

---

From a Book marked "Au. K." p. 202,
29 *Feb.* 1655.

Jonathan Westwood, minister of Croydon, petitioning for arrears, upon the humble petition of Jonathan Westwood, minister of Croydon, in the county of Surrey, thereby charging himself with the sum of 13*l.* in arrears, for tenths, since the 8th of June, 1649; it is ordered, that, upon the said Mr. Westwood his producing a certificate from Mr. Lawrence Steele, treasurer, that he hath received of him the said sum towards the payment of the arrears of the augmentation, the said Mr. Steele by former orders standeth charged to pay him; that thereupon Mr. Allen Nye, receiver, give unto the said Mr. Westwood an acquittance for so much of the tenths of the said vicarage as is in arrear; and that the treasurer do likewise give unto the said Mr. Nye his receipt for so much money by him received of the said Mr. Nye for tenths.

John Thorowgood, Richard Sydenham, John Pocock, Ralph Hall, John Humfrey.

In a Book marked " A. P. N° 9. 29," fol. 212.

Croyden, June 9, 1657.

Whereas these trustees have, by order of the 11 Oct. 1655, granted unto Mr. Jonathan Westwood, minister of Croydon, in the county of Surrey, among other things, the yearly sum of 4*l.* 8*s.* out of the tythes of Bartholomews without Chichester, to be accompted from the 25 March, 1655, and the rest of the said tythes is hitherto in arrear, so that there is due unto the said Mr. Westwood the sum of 8*l.* 16*s.* for two years ending 25 March last, it is ordered that Mr. Lawrence Steele, treasurer, pay the said sum of 8*l.* 16*s.* in full satisfaction of the said arrear.

John Thorowgood, John Humfrey, Ralph Hall, Richard Yonge, John Pococke.

Ibid. fol. 213.

June 9, 1657.

An order of the same trustees, that the sum of 4*l.* 8*s.* per ann. be transferred and charged upon (and from time to time paid to the said Mr. Westwood) out of the rents and profits of the tithes of the sub-deanry of Chichester, to hold for such time as he shall discharge the duty of minister of the said place, untill further orders of these trustees; and that Mr. Allen Nye, receiver, do pay the same to him accordingly. Dated June 9th, 1657.

No. VI.

Commissio ad Vendicand' Clericos Convictos in Villa de Croydon. (*From Strype's* "LIFE OF WHITGIFT," *Appendix, p.* 89.)

JOHANNES, divina Providentia, &c., dilecto nobis in Christo, Samueli Finch, clerico, vicario Vicarie perpetue Ecclesie Parochiali de Croydon in comitat' Surrey, Ecclesie nostre Cathedralis et Metropolit. Christi Cantuar. jurisdictionis immediate &c. salutem et gratiam. Ad vendicand. exigend. recipiend. et examinand. quoscunq. clericos coram quibuscunq. Justiciariis Domine nostre Regine, de super Feloniis quibuscunq. infra Croydon in Com. Surrey Ecclesie nostre Cathedral. et Metropolit. Christi Cantuar. prædict. impetitos, indictatos, sive convictos: Atq. hujusmodi Clericos ad Beneficium et Privilegium clericale in casibus a jure et consuetudine, ac hujus regni Anglie Statutis admissis et pprobatis, recipi et admitti, petend. et requirendum: Cæteraq. omnia et singula alia faciend. exercend. et expediend. que impremissis, et circa ea, necessaria fuerint, seu quomodo libet opportuna, vobis, de quorum fidelitate et circumspectione, et industria in hac parte concedimus conjunctim et divisim committimus vices nostras, et plenam in Domino, tenore presentium, concedimus potestatem. Vosq. commissarios nostros ad omnia et singula premissa exequend. conjunctim et divisim ordinamus, deputamus, et constituimus per presentes. In cujus Testimonium Sigillum nostrum apponi fecimus. Dat. 17 die mensis Julii, 1584, et nostre Translationis primo.

No. VII.

Statutes of Ellis Davy's Almshouse.

Fundatio Hospitalis Pauperum in Villa de Croydon.
(*Ex* " Registro Morton," *fol.* 199 *a.*)

Dated at Croydon, 27 April, 1447.

To all trewe Christen people, these present letters indented seying or hering, Ellis Davy, citizen and mercer of the city of London, sendith greting in our Lord God everlasting. Knowe ye that I the foresaid Ellis Davy, by vertue and auctorite of certain lres patentes of our Sovereign Lord the King that nowe is, bearing date at Westminster, the 25th of December, the year of his reign 23d, by me purchased, have reised up, made, founded, and stablished, in the honor of God, and of the Blessed Virgin Marie, and of St. John Baptist, whose help I have first besought and called to my begynnyne, a perpetual Almes-house of VII pouer people, whereof be called by name a tuter, to dwell perpetually at Croydon in the shire of Surrey; that is to say, in my messuage with the appurtenaunces by me there belded and ordeynid for the same entent, which is called myn almes-hous, and it is settuate bytwene the tenement of William Olyver late vicar of Croydon, which Joh. Ffauxwell there nowe dwellith in, on the south partie, and the ryver there on the north partie.

Also I have put, settee, and ordeigned in the same almes-house VII poure people; that is to say, Pers Stanlock for tuter, Henry Corde, John Christmasse, John Cooke, John Tapcliff, John Shirburne, and Elyn Umfrey, for poure people, with the same tuter, to begyne, take, and have the succession and benefite of the same almes-house and capacite thereof, after the fourme and effect as is conteyned in the said lettres patentes.

Also, by virtue as well of the same lettres patentes as of the lettres patentes of the full Reverent Fader in God John by the grace of God archiebishop of Canterbury, bering date the 17th day of the moneth of Februer, the yere of the incarnacion of our Lorde Jesus Christe 1443, and of our said Sovraigne Lord the King that nowe is 23, and also by the lettres of John by the suffraunce of God abbot of the monastry of Saint Savyor, Barmundsey, in the shire of Surrey, and the Cōvent of the said monastry, beryng date in there Chapter-house the 20th day of the month of December, the yere of the said incarnacion 1445, and of our said Souvraigne Lord the Kynge that now is 24, as for any intresse that they or any of theym hath in the said message by me severally purchased. I the said Ellys have graunted and assigned by these present lettres to the foresaid tuter and poure people, and to his successors for his inhabitation and dwellyng for evermore, of the chief lords of the fee by service thereof due and of right accustomed.

Also I will and ordeyne, that the same house of poure people be called Elys Davie's almes-house of Allimen, from the tyme foreward, and that I for verry founder thereof be holden and taken for evermore. And that the tuter and poure people, and his successors thereof, by the name and under the name of tuter and maister and of poure people of the almes-house of Ellis Davy att Croydon, in the shire of Surrey, mowe complete and be completed; and also al manner of actions, causes, and querellys, real, personall, and mixtis, of what kynde or nature they ben, byfore all manner juges, seculer or spirituell persone, and in the same answer and be answered, and in the same defend, and that they have a comyn seale to serve for the nedis and doyngs of the same almes-hous for evermore.

Also I the said Ellys woll ordeyne that the foresaid poure people and his successors shul be men only, or els men and women togedre, after the good and sadde advise and discrecion of the governours and overseers of the same almes-hous undre-wretyn.

Also I will and ordeigne that A., now vicar of the church of Croydon aforesaid, and also the two chirchwardeines of the same chirch, and iv of the moost worthi men, householders and parishioners, dwellyng and resident within the towne of Croydon, and their successors; that is to say, the vicar and churchwardeyns of the said parish, and iv other of the moost worthy men, householders and parisheners, alwaies resident and abiding within the town of Croydon, be gouvernours of the foresaid almes-hous, and gouvernours of the foresaid almes-hous be called; and that Richard Ritche, John Cotford, John Reynkyn, and John Baron, now masters and wardeyns of the craft or occupation of the mercers of London, and their successours, maisters, or wardeyns of the said crafte or occupation, which for the tyme shall be overseers of the said almes-hous, and overseers of the same almes-hous be called for evermore.

Furthermore, as touching all manner of putting in and owte of such tuter and pour people, and alle other maner, rule, and gouvernaunce of theim, and of the said almes-hous, I reserve it unto mysilf all my lief during; and for the executing and mayntenaunce thereof after my decease, I stabelish and ordeyne, by these present letters, that, as ofte as any avoydaunce, by deth or otherwise, fallith in the said almes-hous, of a tuter, so ofte alwaies, within xx daies next ensuyng after such voydaunce, oon of the said almes-hous convenient and able, if any such be within, and els withoute the said almes-hous of the foresaid town of Croydon; and yf none such convenyent persone to have such a rule and governaunce of tuter can be had, neither in the said almes hous neither towne of Croydon, thenne of some other towne, village, or hamelet nyght abowte the foresaid town of Croydon, so that no one of the same townes, villages, or hamelets, excede the space of iv myle at the moost from the foresaid church of Croydon, by the said governers, that is to say, the said vicar and church-wardeyns, and such iv other moost efficient men of the said towne of Croydon, be preferred and chosin, and ordeyned tuter of the said almes-hous. And yf the foresaid governers be necligent, and not preferre, chose, and ordeyne such a tuter in the fourme aforesaid, within such xx daies sett and lymitted to theym, as byfore is rehersid, then I wol and ordeyne, that after any suche xx daies so necligently overpassid, the preferring, chosing, and ordinaunce, yfe the said tuter to the overseers aforesaid of the said almes-hous shall only belong and perteyne, for to choose and ordeyne of such towns, villages, and hameletts aforesaid, and of noon other.

Provided evermore, that by this ordynaunce no manner prejudice shall growe or be engendered in any wise to the foresaid governors, in other times whenne they be diligent within every such xx daies next suying every voydance of such tuter.

Also, I stablishe and ordeyne, that as oft as any voidaunce of any of the remnant of the poure people being in the foresaid almes house, byside the said tuter and poure people, within xv daies next suying every such voidaunce, chose another unto them of the townes, villages, or hameletts aforesaid, and specially of the said town and parish of Croydon, if any be there. And if it be so that the said tuter and poure people be necligent in there election, so that the said xv daies at any time spevyn there election, not made of such men or wommen so by him chosin, be not in the said almes-house by such xv daies, and that every such time the churchwardeyns of Croydon then for the time being only, and noone other, chose, ordeyne, and sett in the same voidaunce, such as poor man or woman of the towns, villages, or hameletts aforesaid, and of noon other. Provided alwaies, that this ordinaunce be no manner prejudice to the foresaid tuter and poure people in other tymes, as oft as whenne they were diligent and spedefull in there election, within any such xv daies next ensuying eny such voydaunce that happith so for to be beside the foresaid tuter and the election of the same poure people shall be in this manner of fourme, that first they shall name and condescend upon another persone, and which of the said two persons that hath most voyces, that persone having moste voyces shall be admitted and takyn into the said almeshous; and if so be that the voices of the said poure people be eqall of every of the said three tymes, that partie that the voice of the tuter is uppon be take and hadde for the more partie; for the voice of the said tuter in all such cases shall stand for two voices.

Also, I will and ordeyne, that if any other persone than only of the said towne of Croydon, or than of such other towne, village, or hamelett of Croydon, or than of such other town, village, or hamelett nygh abowte, not passing the space of iv myle from the said church of Croydon, be takyn and admitted into the foresaid almes-house, than all such manner of proysion and admission be utterly void, of noo strenght. But thenne I wool and ordeyne, that every such persone that happith so to be admitted into the foresaid almeshous, that was of any other place then only of such town, village, or hammelett above specified, be expelled and put out of the said alms-house by the said overseers, and another person of one of such towne, village, or hamlette aforesaid, in his place by the foresaid overseers be sette in and admitted; for myne entent and wil is, that poure fette people of the said towns, villages, and hamlettes, and specially of the said towne of Croydon, that hath be housholderers or trewe laborers, and dwellid and contynued in some of the said places by seven hoole yeris togeder or more, next byfore such admission made into the said almshouse, and have not whereof to susteyne them with, and may no longer labour for to gett sufficiently livelode, be provided and admitted into the foresaid almshous.

Also, I will and ordeyne, that persone that so shall be provided, and ordeyned, and admitted into the foresaid almeshouse, be a person meek of spirite and chaste of body, and named of good conversacion, and also destitute of temporal goodes by the which he might competently live yf he were noon of the nombre of the said poure people. Wherefore, I beseech and pray, louly and devoutly, and also, as moche as I may with reverence, I charge all thoo to whom parteynith any provision or election of tuter or poor mann or woman by this ordynaunce, that, as they will answer afore the high Judge at the dredeful day of dome, all manner inordinat and singular affection, and corrupcion of favor, praier, or receiving of gefts, be utterly excluded and put away from the harts and handes; and that they only provide and ordeygne able and devout persones of such tuters and poure people as oft as such voidance of theim fallith.

Also, I woll and ordeyne, that the said tuter and every poure persone have a place by himsilf within the said almshous, in the whiche he may ligge and reste, and by himsilf alone, withoute lette of any of his felawes entende to the contemplacion of he will. But I forbede and charge that noo persone of them so being in his owen place, make any noise or disturbaunce, in letting or troubeling any of his felawes, or any of his felawes lette him or distrouble him or them so being in there place or places. But that every of them pesebly and quietly behave himsilf agenst other in such wise, and in all other of wise. Also, I charge theym and their successors, and everiche of them, peasebly and quietly behave himsilf agenst other in such wise, and in all other manner of wise as good is; and to occupy themsilf in praying and in beding, in hering honest talking, or in labours with there bodies and hands in some other occupations, to the laws and worship of Almighti God, and profit to theim and to there said almshous.

Also, I will and ordeyne, that the tuter of the foresaid almshouse that now is, and his successors, tuters of the same almshouse, be bounde by the tenure and auctorite of this ordynaunce for to admitt and receive all manner of poure people by the foresaid churchwardeyns and overseers, and every of theim, in the forme aforesaid to be provided and ordeyned.

Also, I will and ordeyne, that the said tuters and poure people, and there successors, have and take, for his sustenance and lyving, xv l. xii s. yerely of the sum of xviii l. which I will ordeyne, or to be ordeyned to theim yearly for their sustynaunce, and to observe and kepe other chargs hereafter specified; that is to say, the foresaid tuter shall have and take every weke, for his sustenaunce, xiid., and every of the remenaunt of the said pour people shall have and take for his sustenaunce every weke xd.; with the which partis, pencions, and porcions wekely, I will they hold them fully content; and that they beg not, on the payne that hereafter is written and ordeyned agenst such of them as be found corruptible.

Furthermore, I establish and ordeyne, that the said tuter and poure people be bound by the tenure of this present writing, every day in the church of Croydon aforesaid, here all manner divine service there to be songe and saide, and pray specially for the estats of our sovrayne lord the Kinge upon his kneys, iii pater-nosters, iii aves, and a crede, with special and hartily recommendacion of me there said founder, to God and to our blessid lady, maydon Marye, and also everiche of them, tuter and poor people, other tymes of the day, when he may best and commodiously have leisure thereto, or when he or she seith most convenyent tyme, say, for the estate of all the sowlis abovesaid, iii sawters of our lady at the least; that is to say, thrice i ave-maries, with xv pater-nosters, and iii credis; but yf he or she be letted by feblenesse, or other lawfull reasonable cause.

Also, I will and ordeyne, that the said tuter and poure people and there successors, once in the day at the least, in case it may be in the said church of Croidon, where I propose me fully, yf God will, for to be buried, that is to say, after the high masse and whenne compelyn is doon, come togedre about my buriall, and they that caunt shall say for my soule and the soules aforesaid, this psalme de profundis, &c., with the versicles and other orisounes that longith thereto; and they that cannot shall say iii pater-nosters, iii aves, and a crede; and after this doon, the tuter and one of the eldest men of them all shall say, opinly in English, "God have mercy on our founder's soule and on all Christiens!" and they that standith abowte shall answer and say "Amen."

Also, I stablishe and ordeyne, that the foresaid tuter and poure people that now be, and their successours to come, be bounde by the tenure of this ordynaunce to dwelle and abide contynuelly within the foresaid almshous and bounds thereof, as such other poure people in like almshousis and hospitales commenly dwelle, and be bounde to abide and dwelle, and that every day, both at mete and at souper, the ete and be fedde within the said almshouse, onlesse than they be letted by a resonable cause; and whiles they be at mete and souper, I will that they absteyne them from vayne and evill woords in as moche as they may. And yf they will any thinge talke, that it be honest and profitable.

Also, I will and ordeyne, that the over clothing of the said tuter and poure people of the foresaid almeshows be darke and browne of colour, and not staring, neither blasing, and of easy price cloth, according to ther degree.

Also, I stablishe and ordeyne, that neither the said tuter that nowe is, nor any of his successors, tuters of the said almeshows, for to com, absent him in any wise from the said almeshows, by the time of vi daies in all the yere, continuelly and discontinuelly, withoute licence of me while I lief, or of the governers and overseers of the said almeshows after my decease; and that it be than for necessary causis in honest placis, that any of the other poure people of the same almeshows, in every manner of kinde absent theim from the said almshows by on hoole day, or go withoute the bounds of the said almshouse, out of sight from the same almeshows, withowte licence of the said tuter or his successours, yf he be present, or, in his absence, withoute leve of his attorney or depute, lesse then . . . . the gretter need compelle and ask it, or that there be some other cause resonable, which is to be examined and approvid by the same tuter or his debite.

Also, I will and ordeyne, that the foresaid tuter and his successours, as ofte as whenne he shall goo owte into the towne or into any other place, that he and his successours ordeyne oon of his fellawes moost sadde and wise to occupy his occupacion for him till he come ageyne.

Also, I will and ordeyne, that they of the said almeshous be mighte and hoole of body, specially wommen, yf there be any within help, and minister unto ther fellawes of the foresaid almeshous that be sick and felle in there necessaries, as ofte as they have verry neede of help and ber con'.

Also, I will and ordeyne, that the said tuter and poure people have a commyn chiste, in the which chiste they shall putte their commyn seale, and also their charteris, dedis, letters of licence, and privilegs, and this my ordynaunce, and other there myniments and escripts and tresure of there said almeshous, and other things, and which shuld be seem to the foresaid tuter and poure people expedient for the commyn profit of the same almshous; the which chiste I will be putt in a secrete and a seller place within the bounds of the said almeshous, and that to the same chist there be iii divers lokks, and to every lok a key; whereof oon key I woll be in the kepinge of the foresaid tuter, and another key thereof I will be in the kepinge of the eldest felawe of the foresaid almshouse, and the thyrde key I will be in the kepinge of oone of the other felawes of the foresaid almeshouse, every year to be chosen by me while I life, and after my decesse by the said tuter and poure people of the foresaid almeshous; and that no man presume upon him for to hold or keep all the said iii keyes in his owen governaunce at ones, neither with the said commyn seale any thing be ensealed without the licence, consent, and advice of me while I life, and after my deceasse of the foresaid gouvernours and overseers for the tyme being.

Also, I stablish and ordeyne, that all the moneys and tresour of ther commyn goods and rents every yere, after the rekenyng by the tuter made, the which excedith over the expence of the foresaid almeshous, and also the notable jewells of the said almeshows, that be not behovefull every day, shall be kept in the same chiste.

Also, I stablish and ordeyne, that noone of the said tuter and pour people that now be, and shal be in time commyng, presume in any manner or wyse to waste or consume, or give or ley to wedde inordinately, the goodis or any parcel of the goodes belonging to the said hous. But that every of them stody and labour to keepe forth and encrease the said goodes with all his myght; and when any of them shall passe out of this world, that he give or leve all his proper goods to the same almeshous.

Also, I will and ordeyne, that no leper ne madman, nor any other person contynuelly vexed with intollerable seekness, be admitted into the foresaid almshouse; and if it happe that any person that is, or in tyme comyng shall be admitted into the foresaid almshouse, become madd, or woode, or to be enfected with leper or such other intollerable seekness, then I woll and ordeyne that every such persone be putt oute of the said almeshous, leste that infect his felawes, and goo to som other place in the which he may be resceived, where I will that he have evry weke 10đ. for his lyvelode and other necessaries, as he shuld have in the said almshous, and that he be accomptid for one of the said almeshous and of the noumbre of the foresaid vii poure people during his lief.

Also, I stablishe and ordeyne, that in case any of the said poure people, after they be admitted into the foresaid almshous, happen to be promoted or avaunced to any inheritaunce, or otherwise unto the summe of iiii marcs or above by the year, thenne he so promoted or avaunced be putt owte of the same almeshous, and of the nombre of the foresaid vii poure people, and another in his stead and place be resceived and admitted; and if any of the said almeshous, after his admission, be semblable and like inheritaunce, or other casuell advaunced, atteyne or come to any such yearly livelode, that is to say, within the said liiis. ivđ. by yere, then I will and ordeyne, that oone halfe of the same somme evenly, evry yere, without fraude or mall engin, be put and kepte into the foresaid chiste, to the commyn profitt of the same almeshous, and that other halfe of the same somme shall go to the foresaid persone so promoted or avaunced, with his own portion of the said almeshous before assigned, with the which I woll that he holds him content; and in case that he will not obeye, but contrarie this ordynaunce, than I will that he be put owte of the foresaid almeshouse utterly, and another person be provided and put in his place.

Also, I stablish and ordeyne, that the tuter take hede and see that the goodes of the foresaid almeshouse, which shall come in any wise to his hands, be well and trewly ministrid; and yf any of them have to be desceveryd, to gadre them togedre, and all suche goodes so by his labour gaderid togedre to kepe trewly to the awaile and use of the said almeshous, and to do all the husbandry of the same almshous in as moche as he may do goodly; and also enforce himsilf to edifie and norrish charite and pease amongs his felawes, and shewe with all his besiness, both in worde and in dede, ensamples of clenness and of vertue; unto which tuter I charge and ordeyne, that all the remmenaunt of poure people of the said almeshous that now be, and their successours, in all things that towchith the good gouvernaunce, and honeste, and profitt of the same almeshous, give due obedience and attendaunce, as right askith.

Also, I will and ordeyne, that evry tuter of the said almeshous, within a month after he is admitted, take to him two of his felawes moost discrete by the denominacion of the foresaid governers or overseers, and make a just inventarye of all the commyn moveable goodes of the said almeshouse, and the inventarye so made, withoute tarrying or delay, shew or putt up unto the foresaid gouvernours or overseers; and in the ende of evry yere, or whenne any tuter cessith his office and charge, the new tuter shall do in the same wyse, and byfore the said gouvernours or overseers yeld and make a just accompt of alle the tyme of his administracion, that alle men may opinly knowe in what a state evry tuter hath resceived, kept, and left the foresaid almeshouse for his tyme.

Also, I will and ordeyne that noone of the said poure people which is lower in degree then the foresaid tuter, lye not owte of the foresaid almeshous by nyght, in the foresaid towne of Croydon or elsewhere, withoute a resonable cause, to be discussid and examinyd by the jugement of the said tuter, and also of the foresaid poure people, be custumably dronkley glotons or rigours amongs his felawes, or haunting taverns, or be unchast of his body, or walking or gazing in the opyn stretis of the said towne of Croydon, by day or by night, oute of sight from the said almshouse, but if it be only to the church and churchyarde of the same, with a reasonable cause, to be pondered and discussid by the discrecion and jugement of the said tuter, or of the foresaid gouvernours or overseers; and what persone of the said poure people of any of the said defaultis, or of any such other defaultis, or vice like to these, be openly defamed and notably marked, I wool that he be under no . . . . and correctid by the said tuter, by twise withdrawing his portion, more or lesse, after the discrecion of the foresaid tuter; which portion so withdrawen shall been applied and put into the commyn chiste of the foresaide house. And if any of the saide poure people be warned, rebuked, and punished, for such defaults and vices as afore is rehersid, or other like to them, be founde the iiid tyme defected thereof, after my deceasse, afore the said tuter, or afore the said gouvernours or overseers, thanne I woll and ordeyne, that eny such misgovernd persone so iii tymes founde defected to be held and taken for their incorrectible and intollerable persone, and utterly be put away by the said tuters, and gouvernours and overseers, from the said almeshous, and from all manner of benefitte, profitt, and advantage, that he should have had therein yf he hadde be well governed; and that another convenient persone be provided and sette in his place, in the manner and fourme afore expressed.

Also, I will and ordeyne, that, yf any of the said poure people afore the saide tuter, gouvernours, or overseers, after my deceasse, be opinly dislaunderer and noisy, and after their discrecion and consideracion be resonable convicte for a commyn distroyer or an inordinat waster of the goodes of the foresaid almeshous, or for an opyn lechour or avowter, or for a doer of more grete synnes thenne thise above expressid, thanne I woll and ordeyne, that eny such persone, at the first tyme that thus he is convicte, be utterly put away from the said almeshous; and in caas that any such persone, so put oute after my decesse, will complayne or appele, that it be only to the overseers of the foresaid almeshous, and to noon other.

Also, I will and ordeyne, that the defautis and trespasis of the foresaid tuter for the tyme being, aftre my deceasse, be refourmed, correctid, and punishid, by the gouvernours and overseers of the foresaid almeshous, under this fourme, that is to say, as well be the withdrawing of his portion and part more or lesse, after the quantite of his trespace, afore the consideracion and disposicion of the gouverners and overseers to be mesured and taxed, as by put oute and removing of the said tuter from his office, pencion, and place, which he hath in the foresaid almeshous, yf his offence and gilt askith it.

Also, I wool and ordeyne, that the gouverners and overseers of the foresaid almeshous, and ther successours, have poure for to make, after my deceasse, all manner statutes and ordinaunces y[t] may be seem to theim to the encreace of the said almeshous, and wellfare of the foresaid tuter and poure people, and of ther successours; and that all such statutis and ordinaunces, so by the same gouverners or overseers or her successours made, be observed and kept of the said tuter and poure people, and ther successours, and everiche of them, so alwaies that suche statuts and ordinaunces be according to resone, and not contrary to thise my statuts and ordinaunces, or to any other by me hereaftre to be made: And in caas any doubte, ambiguyte, or worse, befalle in mysundrestanding of thise said statuts and ordinaunces, or of any other hereafter to be made by me, thenne I wool that all such statuts and ordinaunces so mysundrestanding be declared, correctid, reformed, and amended, only by the foresaide overseers for the tyme being, or ther successours.

Furthermore, I stablish and ordeyne, that the foresaid tuter and poure people, and ther successours, hold and make evry yere, aftre my decesse, in the church of Croydon aforesaid, my yeris mynde solempnly, with full service by note usid and accustumid in exequyes of mortuarys, begynning twelvemonth after the day of my deceasse, and so aftre that yerely on such alwaies as God wil that I deceasse, for evermore, yf I decesse in suche tyme of the yere that the same day may be kept, or els as sone aftre as hooly church will suffre it to be doon. Iwch yeris mynde I will that the foresaid tuter and poure people pay yerely of the foresaid xviii l. to theim by me to be ordeyned, unto the vicar of the church, yf he be present at the said yeris mynde, xx d.; and to either of the said ii church wardeyns, yf they be present, xx d.; and to every priest and parish-clerk of the same church, iiii d.; and that the said tuter offer i d., and every of the said other people offre ob. at the masse of requiem of the same yeris mynd.

Also, I woll and ordeyne, that the foresaid tuter and poure people be bound by this present ordinaunce to pay evry yere unto the iiii maisters or wardeyns aforesaid, and to her successours, overseers, and visitours for the same almeshous, so that ii at the least of the same iiii maisters or wardeins which for the time shall be com evry yere to the foresaid almeshous, and see and visite that the same house and also the tuter and poure people thereof be well gouvernyd, and such defaults as by fore theim be founde, correcte, punissh, and amende, as moch as in theim is, for to doom xxvi s. viii d., that is, for evry of them vi s. viii d., and also xiii s. iiii d. for the costs of the same ii maisters or wardeyns, spended in their commyng thither aforesaid evermore; that the said iiii wardeyns, evry such yere as ii at the lest of theym com not and visit the foresaid almeshouse and poure people, in the fourme aforesaid, shall no thynge have of the foresaide xxvi s̃. viii d̃., neither of the said xiii s̃. iiii d̃. for their costs: But thenne evry such yere the foresaid xxvi s̃. viii d̃. for their labours, and the foresaid xiii s̃. iiii d̃. for their costs, remayne stille with the same tuter and poure people, and her successours, in there said commyn chiste, to the encrease and availe of the same hous, among their commyn tresoure.

Furthermore, I ordeyne and faithfully stablish, that if man or woman, in my foresaid Elis Davys almeshous, begge or aske any silver, or else any other good, within the said almeshouse, or else withoute, in any other place, that thenne I wool fully that all suche be expellid and put oute at the first warnyng, and never be of the fellowship, neither never more to come into the said almeshous.

Also furthermore, I the said Elys have given and graunted unto the foresaid tuter and poure people, and to there successours for evermore, iii cotagis, with the gardines sett in Croydon aforesaid, betwene the said ryver on the south partie and the king's highway on the east and north parties, and the gardynes sumtyme blaunchards on the west partie; saving alwaies I woll and ordeyne that the same iii cotagis, and also another cotage, parcel of the said messuage above givin and graunted, situate by the south side of the mansion of the said almeshous, be sette oute yerely to ferme for the moost availe that may be, and that all the issues and profitts comyng of the ferme of the said iii cotages be occupied and turned to noon other use than only to the reparacion of the same almeshous and cotagis; and whatsoever of the same issues and profitts that leveth over, I will that it be put into the foresaid chiste, there saufly to be kept to thencrease of a commyn tresour for the said tuter and poure people; so, if the said almeshous and iiii cotagis, or any of them, shuld nede to be new bilded, therewith to bild it, or other comyn nedis of ther foresaid almeshous to susteyne and do.

Moreover, I will and ordeyne, that this my present fundacion and ordinaunce, and all and everiche chaptre of the fundacion and ordinaunce, be redde opinly and clerely expowned evry quarter of the yere, once at the lest, before the said tuter and poure people, and her successours; and the same tuter and poure people have within theimsilfe, in the said almeshous, a copy of the said statuts, so that whenne they wool they may rede the chapters of this ordinaunce, and the better have theim in mynde: Saving alwaies, I reserve unto mysilfe full poure, by thise my present letters, of alle manner of things that parteyneth to the said almeshous, tuter, and poure people, as long as I life, to chaunge these foresaid ordynaunces yf nede be, or to correct or dispence and make new statuts, and to revoke thoo that nowe be made, in case I see it expedient, this my present ordinaunce notwithstandinge. Nevertheless, as for nowe this tyme I pray and hertely beseche the said tuter and poure people that nowe be, and shal be in tyme comyng, that they love togedre and kepe charite amonge theim, and serve God, and pray hertily for the soules beforerehersed, at this my present ordinaunce will, and they be conversaunt and love togedres, in such wise in this foresaid almshows, and that after thende of this life, they may comme to the hous of the kyngdome of hevin, the which to poure people is promised by the mouth of our Lord God. Amen.

In witnes of all the whiche things above expressid, to that oon part of this present writing indented, with the foresaid tuter and poure people and her successours for evermore remayning, I the said Elys have sette my seal, and we the foresaid tuter and poure people of the said almeshous, being right glad and joyfull of the good grace of our sovraigne lord the kyng, and of the foresaid reverent fader in God the archebushop of Caunterbury, and the abbot and convent of the monastery of Berdmonsey, and of the foresaid Ellis Davy our founder, and of the good, diligent, and effectual labours of the same our founder, and yeving thankis and loving and praising thereof to Almyghti God, and to the blessid virgin Marye, his moder, Saint John Baptist, and to all the company of hevin; and taking upon us the corporacion and succession yevin and graunted by the foresaid letters patent of our saide souveraigne lord the king, and also the benefite and charge of this present fundacion articuled, for to fulfill and do it in all and evry article in manner and fourme as above is declared, to that other part of this present writing indentid with the foresaid Elys our foundere while he lyvith, and after his decesse with the foresaid thenne our gouverners and overseers of our said almeshous and ther successours evermore remaynyng, have sette our commyn seale.

> Yevin at Croydon aforesaid, the xxviith day of the moneth of Aprill, the yere of the incarnacion of our Lord Jhu Christ, MCCCCXLVII, and of our sovraigne lord the kynge Henry the vith, after the conquest of Engeland the xxvth.

No. VIII.

Letters Patent for building the Hospital of the Holy Trinity at Croydon, with Licence of Mortmain and incorporating the same, 22nd Nov. 30 Eliz.

ELIZABETH, Dei gratia, Angliæ, Franciæ, et Hiberniæ Regina, Fidei Defensor, &c. Omnibus ad quos præsentes literæ pervenerint salutem. Cum reverendissimus in Christo Pater prædilectus consiliarius noster Johannes Whitegifte Cantuar' Archiepiscopus, totius Angliæ primas et metropolitanus, perpendens mortales omnes in hoc mundo positos ut Dei omnipotentis gloriam illustrent, ac humano generi quam maxime prosint, illamque maximam beneficentiam merito censendam esse, non quæ fluxæ istius ac labilis humanæ vitæ angustis terminis continetur, sed quæ in miseros et egenos Christianos, ipsius Jesu Christi Salvatoris nostri mystica membra, ad diuturnitatem per omnes succedentium sæculorum ætates (modo Deo ita videatur) propagari utiliter poterit; Ptochodochium quoddam sive Hospitale pauperum intra villam de Croydon in comitatu nostro Surr', de mundanis suis facultatibus quas Deus ei tanquam dispensatori concedidit, fundare, exigere, dotare, et in perpetuum stabilire statuerit, nostro prius regio assensu ad id exhibito, quem humiliter ac dimisse a nobis petierit: Sciatis igitur quod nos, tam laudabili pioque ipsius instituto faventes, idemque summopere promovere cupientes, de gratia nostra speciali ac ex certa scientia et mero motu nostris, volumus, concedimus, et ordinamus, pro nobis, hæredibus et successoribus nostris, quod de cætero in perpetuum sit et erit unum Ptochodochium sive hospitale pauperum in Croydon prædicto, pro sustentatione sive relevamine pauperum et indigentium quorundam Christianorum, perpetuis temporibus duratur'; quod quidem hospitale vocabitur Hospitale Sanctæ Trinitatis in Croydon ex fundatione Johannis Whitegifte Cantuariensis Archiepiscopi; et hospitale illud, per nomen Hospitalis Sanctæ Trinitatis in Croydon ex fundatione Johannis Whitegifte Cantuariensis Archiepiscopi, erigimus, cedamus, fundamus, et stabilimus firmiter per præsentes; et quod hospitale illud de cætero sit et erit aut possit esse de uno custode et pauperibus per eundem Johannem Whitegifte Archiepiscopum et successores suos secundum ordinationem in his literis nostris patentibus specificat' eligend'. Et ut intentio ac propositum hoc pium meliorem firmioremque sortiatur effectum; atque ut bona, terræ, tenementa, redditus, reventiones, et alia hæreditamenta, ad sustentationem hospitalis prædicti ac custodis pauperumque et aliorum in eodem degentium, posthac concedend', assignand', et destinand', melius gubernentur, tractentur, regantur, et insumantur, pro perpetua continuatione ejusdem; Volumus, concedimus, et ordinamus, pro nobis, hæredibus et successoribus nostris, per præsentes, quod de cætero sit et erit in perpetuum unus custos dicti hospitalis Sanctæ Trinitatis in Croydon prædicto, ac terrarum, tenementorum, redditunm, reventionum, possessionum, aliorumque hæreditamentorum ejusdem hospitalis, necnon bonorum et catallorum ejusdem; qui erit et vocabitur Custos Hospitalis Sanctæ Trinitatis in Croydon ex fundatione Johannis Whitegifte Cantuariensis Archiepiscopi; quodque de cætero perpetuis futuris temporibus sint et erunt sex, septem, octo, novem, decem, aut aliquis alius numerus sub numero quadraginta pauperum in eodem hospitali sustentand', manutenend', et relevand', ac per prædictum Johannem Whitegifte Cantuariensis Archiepiscocum et successores suos, secundum intentionem harum literarum nostrarum patentium, de tempore in tempus eligend', nominand', et assignand', qui similiter erunt et vocabuntur Pauperes Hospitalis Sanctæ Trinitatis in Croydon ex fundatione Johannis Whitegifte Cantuariensis Archiepiscopi. Et ad munus et officium custodis hospitalis prædicti bene et fideliter exequend' et exercend', nos, ad humilem petitionem dicti Johannis Whitegifte Cantuariensis Archiepiscopi, pro prima vice elegimus, nominavimus, assignavimus, et constituimus, ac per præsentes eligimus, nominamus, assignamus, et constituimus dilectum nobis Philippum Jenkins fore et esse primum et modernum custodem dicti hospitalis, ac terrarum, tenementorum, reddit', reventionum, possessionum, aliorumque hæreditamentorum ejusdem hospitalis, necnon bonorum et catallorum ejusdem; ac, ad consimilem petitionem dicti Johannis Whitegifte Cantuariensis Archiepiscopi elegimus, nominavimus, assignavimus, et constituimus, ac per præsentes eligimus, nominamus, assignamus, et constituimus, Johannem Holland, Christoferum Fenner, Reginaldum Scroobie, Thomam Whitehead, Richardum Dibble, et Robertum Curtis, fore et esse primos et modernos pauperes hospitalis prædicti, continuandos in eodem hospitali durant' vitis suis naturalibus, nisi pro aliqua causa rationabili per Archiepiscopum Cantuariensem pro tempore existent' aut successores suos amovebuntur, aut eorum aliquis amovebitur. Et ulterius, de ampliori gratia nostra speciali ac ex certa scientia et mero motu nostris, volumus, ac per præsentes pro nobis, hæredibus et successoribus nostris, concedimus et ordinamus, quod iidem custos et pauperes hospitalis prædicti et successores ipsorum de cætero in perpetuum sint et erunt unus corpus corporatum et politicum de se, in re, facto, et nomine, per nomen Custodis et Pauperum Hospitalis Sanctæ Trinitatis in Croydon ex fundatione Johannis Whitegifte Cantuariensis Archiepiscopi; ac ipsos custodem et pauperes et successores suos per nomen Custodis et Pauperum Hospitalis Sanctæ Trinitatis in Croydon ex fundatione Johannis Whitegifte Cantuariensis Archiepiscopi incorporamus, ac unum corpus corporatum et politicum per idem nomen in perpetuum duratur' realiter et ad plenum, pro nobis, hæredibus et successoribus nostris, erigimus, creamus, ordinamus, facimus, constituimus, et stabilimus firmiter per præsentes; quodque per idem nomen Custodis et Pauperum Hospitalis Sanctæ

Trinitatis in Croydon ex fundatione Johannis Whitegifte Cantuariensis Archiepiscopi, perpetuis futuris temporibus, vocabuntur, appellabuntur, et nominabuntur, habeantque successionem perpetuam; et quod per idem nomen sint et erunt personæ habiles, aptæ, et in lege capaces, ad perquirendum, recipiendum, habendum, et possidendum tam bona et catalla, quam maneria, terras, tenementa, prata, pascua, pastur', redditus, reventiones, et alia hæreditamenta quæcunque, sibi et successoribus suis, in perpetuum seu aliter, tam de nobis, hæredibus vel successoribus nostris, quam de præfato Johanne Whitegifte Cantuariensis Archiepiscopo, hæredibus vel assignatis suis, vel ab aliqua alia persona, seu de aliquibus aliis personis, ad sustentationem, manutentionem, et relevamen hospitalis prædicti, ac custodis atque pauperum ibidem de tempore in tempus degentium et sustentand'. Volumus etiam, et pro nobis, hæredibus et successoribus nostris, per præsentes concedimus, præfato custodi et pauperibus dicti hospitalis et successoribus suis, quod ipsi de cætero in perpetuum habeant commune sigillum ad negotia sua et quamlibet seu aliquam inde parcellam tangent' seu concernent' deservitur'; et quod prædictus custos et pauperes hospitalis prædicti et successores sui, per nomen Custodis et Pauperum Hospitalis Sanctæ Trinitatis in Croydon ex fundatione Johannis Whitegifte Cantuariensis Archiepiscopi, placitare et implacitari, prosequi, defendere et defendi, respondere et responderi possint et valeant, in omnibus et singulis causis, querelis, sectis, et actionibus quibuscunque, cujuscunque generis sive naturæ fuerint, in quibuscunque locis et curiis nostris, hæredum vel successorum nostrorum, ac in placeis, locis, et curiis aliorum quorumcunque, ac coram quibuscunque judicibus aut justiciariis intra hoc regnum Angliæ aut alibi; et ad ea, ac ad omnia et singula alia faciend', agend', et exigend', eodem modo prout cæteri ligei nostri personæ habiles et in lege capaces intra idem regnum nostrum Angliæ faciunt, aut facere valeant seu posse, in curiis, locis, et placeis prædictis, et coram justiciariis et judicibus supradictis. Volumus etiam, ac ex uberiori gratia nostra speciali ac ex certa scientia et mero motu nostris, pro nobis, hæredibus et successoribus nostris, concedimus, quod quandocunque contigerit prædictum custodem hospitalis prædicti ab hac vita discedere, vel a loco et officio suo custodis hospitalis prædicti amoveri, quod tunc et totiens bene liceat et licebit Archiepiscopo Cantuariensi pro tempore existent', et successoribus suis, unam sufficientem et idoneam personam in locum sive officium hujusmodi custodis sic morientis vel amoti eligere et præficere; quæ quidem persona, sic ut præfertur ad officium et locum custodis hospitalis prædicti electa et præfecta, erit et continuabit custos hospitalis prædicti durante vita sua naturali, nisi interim pro aliqua causa rationabili ab officio et loco prædictis amovebitur. Volumus etiam, et per præsentes pro nobis, hæredibus et successoribus nostris, concedimus, quod quandocunque contigerit aliquem vel aliquos prædictorum modernorum pauperum hospitalis prædicti qui nunc sunt, aut qui in posterum juxta formam et effectum præsentium per prædictum Johannem Whitegifte Cantuariensis Archiepiscopum aut successores suos electi, nominati seu assignati erunt, ab hac vita discedere, vel ab hospitali prædicto pro aliqua causa rationabili amoveri, quod tunc et totiens bene liceat et licebit Archiepiscopo Cantuariensi pro tempore existent' et successoribus suis, aliquam personam vel alias personas pauperes et indigentes in hospitali prædicto manutenend' et sustinand' eligere et præficere, et sic totiens quotiens casus sic acciderit. Volumus etiam, ac per præsentes pro nobis, hæredibus et successoribus nostris, concedimus, quod tam idem Johannes Whitegifte Cantuariensis Archiepiscopus de tempore in tempus durante vita sua naturali, quam etiam alius vel alii per istum in ultima sua voluntate in scripto, aut per factum suum manu sua subscripta et sigillo suo signatum, nominandum atque assignandum, faciat et faciant, et facere valeat et valeant et possint, idonea et salubria statuta et ordinationes in scripto concernen' tam veram religionem ac divina servitia infra hospitale predict' de tempore in tempus in honorem Omnipotentis Dei celebranda, quam gubernationem, electionem, expulsionem, punitionem, et directionem custodis et pauperum hospitalis prædicti pro tempore existent', necnon stipendia et salaria eorundem custodis et pauperum, et alia quæcunque idem hospitale seu custodem et pauperes predictos, ac ordinationes, preservationes, et dispositiones possessionum, reddituum, reversionum, aliorumque hæreditamentorum, necnon bonorum et catallorum ejusdem hospitalis, ad sustentationem ejusdem hospitalis concedend', assignand', vel destinand', quovismodo tangent' et concernend': Quæ quidem statuta et ordinationes sic, ut præfertur, faciend', volumus et concedimus, ac per præsentes, pro nobis, hæredibus et successoribus nostris, præcipimus inviolabiliter observari de tempore in tempus in perpetuum: Ita tamen quod statuta et ordinationes prædict', sic ut præfertur fiend', nec eorum aliqua, non sint contraria nec repugnand' legibus et statutis regni nostri Angliæ. Et ulterius, de uberiori gratia nostra, ac ex certa scientia et mero motu nostris, per præsentes damus et concedimus præfato custodi et pauperibus Hospitalis Sanctæ Trinitatis in Croydon ex fundatione prædicti Johannis Whitegifte Cantuariensis Archiepiscopi, et successoribus eorum, licentiam specialem, liberamque et licitam facultatem, potestatem, et autoritatem, perquirendi, recipiendi, obtinendi, gaudiendi, possidendi, et habendi, eis et eorum successoribus in perpetuum, ad perpetuam sustentationem et manutentionem hospitalis prædicti, tam de nobis, hæredibus et successoribus nostris, quam de præfato Johanne Whitegifte Cantuariensis Archiepiscopo, hæredibus, executoribus, vel assignatis suis, vel de aliis quibuscunque personis, maneria, messuagia, terras, tenementa, rectorias, decimas, et alia hæreditamenta quæcumque intra regnum nostrum Angliæ, quæ in toto non excedunt clarum annuum valorem ducentarum librarum, et quæ non tenementa de nobis, hæredibus vel successoribus nostris immediate in capite, nec aliter de nobis, hæredibus vel successoribus nostris, seu de aliquo alio, per servitium militare. Damus etiam, et pro nobis, hæredibus et successoribus nostris, per præsentes concedimus, omnibus et singulis personis quibuscunque, et eorum cuilibet, licentiam specialem, et plenam, liberam et licitam facultatem et authoritatem, quod ipsi et eorum quilibet maneria, messuagia, terras, tenementa, rectorias, decimas, et alia hæreditamenta quæcumque intra regnum Angliæ, non excedend' in toto clarum annuum valorem ducentarum librarum, et quæ non teneantur de nobis, hæredibus vel successoribus nostris immediate in capite, nec aliter de nobis, hæredibus vel successoribus nostris, seu de aliquo alio, per servitium militare, præfatis custodi et pauperibus hospitalis prædicti et successoribus suis in perpetuum dare, legare, concedere, vendere, et alienare possit vel possint, sine aliquo brevi de Ad quod damnum, aut aliqua inquisitione in hac parte habenda seu facienda, statuto de terris et tenementis ad manum mortuam non ponendis, aut aliquo alio statuto, actu, ordinatione, sive provisione, aut aliqua alia re, causa, vel materia quacunque in aliquo non obstante. Ac ulterius volumus, ac per præsentes præcipimus et ordinamus, quod omnia proficua, exitus, et reventiones omnium hujusmodi terrarum, tenementorum, hæreditamentorum, et possessionum, in posterum ad perpetuam sustentationem et manutentionem prædicti hospitalis dandorum et assignandorum, convertantur, disponantur, et expendantur ad sustentationem custodis et pauperum hospitalis prædicti pro tempore existen', ac aliorum officiariorum et ministrorum ejusdem pro tempore existent', juxta ordinationes et statuta per prædictum Johannem Whitegifte Cantuariensis Archiepiscopum, seu per aliquem alium aut aliquos alios per ipsum nominand' aut assignand', prout præfertur, faciend'; necnon ad sustentationes, manutentiones, et reparationes domorum, terrarum, et possessionum hujusmodi, secundum eadem statuta et ordinationes sic ut præfertur faciend', et non aliter, nec ad aliquos alios usus aut intentiones. Volumus etiam, ac per præsentes, pro nobis, hæredibus et successoribus nostris, concedimus præfato Johanni Whitegifte Cantuariensis Archiepiscopo, quod habeat et habebit has literas nostras patentes sub nostro sigillo nostro Angliæ debito modo factas et sigillatas absque fine, seu feodo magno vel parvo, nobis hanaperio nostro seu alibi ad usum nostrum proinde quoquo modo reddendo, solvendo, vel faciendo; eo quod expressa mentio de vero valore annuo vel de rectitudine præmissorum, sive eorum alicujus, aut de aliis donis sive concessionibus per nos seu per aliquem progenitorum nostrorum præfato Johanni Whitegifte Cantuariensis Archiepiscopo ante hæc tempora factis in præsentibus minime factis existit; aut aliquo statuto, actu, ordinatione, provisione, sive restrictione in contrarium inde antehac habit', fact', edit', ordinat', sive provis', aut aliqua alia re, causa, vel materia quacunque in aliquo non obstante. In cujus rei testimonium has literas nostras fieri fecimus patentes. Teste meipsa, apud Westmonasterium, vicesimo secundo die Novembris, anno regni nostri tricesimo octavo.

Per Breve de Privato Sigillo.

KEMPE.

Indorsed—

A Graunte to the Archbishop of Canterbury, for erecting of an Almeshouse, at Croydon in the Countie of Surr'. KEMPE.

Character' per WILL'UM BRENT.

No. IX.

Archbishop Whitgift's Deed of Foundation.

To all true Christian people to whome theis presents shall come, John Whitegift, Archbishop of Canterbury, Primate of all England, and Metropolitan, sendeth greeting in our Lord God everlasting. Whereas, in the session of Parliament houlden at Westminster in the nyne and thirtith yeare of the reigne of our Sovereigne Ladie Elizabeth the Queenes Majesty that nowe is, one acte was made, entituled "An acte for erecting of hospitalls or abidinge and working houses for the poore:" Now this present deede witnesseth, that the said John Whitegift, Archbishop of Canterburie, now being seised of an estate in fee simple in his owne right, and to his owne use, of and in one building of brick, or brickhouse, newly and lately by him built and erected in Croydon, in the countie of Surrey, and of and in certen other houses, gardyns, orchardes, courtes, yerds, and grounds thereunto adjoyning, scituat and being in Croydon aforesaide, doth by the power, vertue, strength, licence, and authoritie of the said acte, by this his present deede to be enrolled in the high courte of Chauncerie, erect, founde, and establish the saide building of brick, or brickhouse, and the saide houses, gardens, orchardes, courtes, yerdes, and groundes thereunto adjoyning, to be an hospitall and abiding-place for the finding, sustentation, and relief of certen maymed, poore, needie, or impotent people, to have continuance for ever; which hospitall, and the persons therein to be placed, the said John Whitegift, Archbishop of Canterburie, hath assigned, limited, and appoynted, and hereby doth assigne, limit, and appointe, to be incorporated, named, and called, by the name of The Hospitall of the Holye Trinitie, in Croydon, of the foundation of John Whitegift, Archbishop of Canterbury; and the same Hospitall, by the name of The Hospitall of the Holye Trinitie, in Croydon, of the foundation of John Whitegift, Archbishop of Canterbury, doth, by force and vertue of the said acte of Parliament, and by this his deede, erect, founde, and establish firmely to have contynuance for ever; and doth also ordeyne, limite, and appointe, that the same hospitall shall consist of one wardeine, which shalbe the headd of the said hospitall; and of maymed, poore, or impotent persons, not exceeding in all the nomber of forty, which shalbe the bodye and members of the said hospitall, and they from tyme to tyme to be chosen, nominated, placed, appoynted, and assigned, according to the true intent and meaning hereafter in theis presents expressed or mentioned. And to the end that the said intent and purpose of the said Archbishop of Canterbury may take the better and more sure effect, and that the landes, tenements, rents, revenewes, and other hereditaments, and also all and singuler goodes and chattells, now or hereafter to be geven, graunted, assigned, or appoynted, to or for the sustenaunce or mayntenaunce of the said hospitall, and of the wardeyne and maymed poore or impotent persons therein for the tyme being abiding, may the better be mayntened, governed, disposed, ruled, and bestowed for ever hereafter; the said Archbishop of Canterbury appoynteth, assigneth, lymiteth, and ordeyneth, by theis presents, that from henceforth for ever there be and shalbe one wardeine of the said hospitall of the Holie Trinitie, in Croydon, of the foundation of John Whitegift, Archbishop of Canterbury, and of the landes, tenements, rents, revenewes, possessions, and other hereditaments of the same hospitall, and also of the goodes and chattels of the same, which shalbe called the Wardeine of the hospitall of the Holy Trinitie, in Croydon, of the foundation of John Whitegift, Archbishop of Canterbury, and that for ever there bee and shalbe maymed, poore, or impotent persons, not exceeding the number of fortie, which shalbe susteyned, maynteyned, and relieved in the same hospitall, and from tyme to tyme be chosen, nominated, placed, appoynted, and assigned, according to the true intent and meaning hereafter in theis presents expressed or mentioned, which likewise shalbe called The Poore of the Hospitall of the Holie Trinitie, in Croydon, of the foundation of John Whitegift, Archbishop of Canterbury: And for the office and function of the wardeine of the same hospitall well and truely to be executed and exercised, the said Archbishop of Canterbury, for the first tyme, hath chosen, assigned, and appointed, and by theis presents doth choose, assigne, and appoint Philip Jenkins to be the first and present wardeine of the same hospitall, and of the landes, tenements, rents, revenewes, possessions, and other hereditaments of the same hospitall, and also of the goodes and chattells of the same: And allso the said Archbishop of Canterbury hathe chosen, assigned, and appointed, and by theis presents doth choose, assigne, and appoint John Holland, Christopher Fenner, Reynold Scroobie, Richard Deeble, Robert Curtis, Edward Holloway, Edward Pringle, Augustine Willis, Robert King, Henry Jefferey, Henry Leaver, and Thomas Elton, to be twelve of the first of the saide maymed poore or impotent persons, not exceeding the number of fortie, of the same hospitall, to contynue in the same hospitall, with the residue of the said maymed poore or impotent persons, not exceeding the number of fortie, hereafter frome tyme to tyme to be chosen, nominated, placed, appointed, and assigned, by the said archbishop, his heires, executors, or assignes, according to the true intent and meaning hereafter in theis presents expressed or mentioned. And further, the said Archbishop of Canterbury doth by theis presents (by force and vertue of the said acte) graunt, ordeyne, lymitt, assigne, and appointe, that the warden and maymed poore or impotent of the same hospitall as is aforesaid, and their successors for ever, be and shalbe one bodie corporate and politique of it selfe in deed and name, by the name of The Wardein and Poore of the Hospitall of the Holy Trinitie, in Croydon, of the foundation of John Whitegift, Archbishop of Canterbury; and the same wardein and poore, and their successors, by the said name of The Wardein and Poore of the Hospitall of the Holy Trinitie, in Croydon, of the foundation of John Whitegift, Archbishop of Canterbury, doth by theis presents, according to the said power and authority before mentioned, incorporate, creat, and make one bodie corporate and politique, by the same name for ever, to the pleasure of Almightie God to endure; and also really and fully, for him and his heires, doth erect, creat, ordeyne, make, constitute, and establish firmely by theis presents; and that by the same name of The Wardeyn and Poore of the Hospitall of the Holy Trinitie, in Croydon, of the foundation of John Whitegift, Archbishop of Canterbury, at all tymes hereafter, they shalbe called and nominated, and by that name have a perpetuall contynuaunce and succession for ever; and that by the same name they be and shalbe persons hable, apt, and capable in the lawe, to purchase, receave, have, and possesse, aswell goodes and chattells, as mannors, landes, tenements, rents, and hereditaments whatsoever (not exceeding the value of two hundreth pounds by the yeare) to them and their successors for ever, aswell of our Soveraigne Ladie the Queenes Majestie, her heires and successors, as of the said Archbishop of Canterbury, his heires and assignes, or of any other person or persons, for the sustentation, mayntenaunce, and relief of the said hospitall, and of the said wardein and poore therin from tyme to tyme abiding, and to be relieved, according to the forme, effect, and true meaning of the said acte of Parliament, to all intents and purposes: And also, the said Archbishop graunteth, assigneth, and appointeth, by theis presents, to the said wardein and poore of the hospitall of the Holy Trinitie, in Croydon, of the foundation of John Whitegift, Archbishop of Canterbury, and to their successors, that they for ever hereafter shall and maye have and enjoye a common-seale, to serve for their affayres concerning the said hospitall, and everie of them, engraved with the historie of Dives and Lazarus, and a scutcheon of the armes of the said John Whitegift, and circumscribed with theis wordes, "Sigillum Hospitalis Sanctæ Trinitatis in Villa de Croydon;" and that the said wardein and poore of the said hospitall, and their successors, by the name of The Wardein and Poore of the Hospitall of the Holy Trinitie, in Croydon, of the foundation of John Whitegift, Archbishop of Canterbury, may and shalbe able to plead and be impleaded, sue and to be sued, defend and to be defended, aunswer and to be aunswered unto, in all and singuler causes, quarrells, suites, and actions whatsoever, of whatsoever kinde or nature they be, in whatsoever places or courts of our said Soveraigne Lady the Queene, her heires or successors, or in the courtes and places of any other person or persons whatsoever, and before any judges or justices whatsoever, within this realme of England, or elsewhere; and to doe and execute, performe and accomplish, all and singuler other things whatsoever, and that as fully and freely, and in as large and ample manner and forme, as persons incorporate, or any other the liege people of our said Soveraigne Lady the Queene, being persons able and capable in the lawe, may lawfully doe and execute in any parte or place within the realme of England; and that the same hospitall, and the warden and maymed poore or impotent persons of the same for the tyme being, and everie of them, shalbe from tyme to tyme ordered, directed, and visited, placed, or upon just cause displaced and amoved, by suche person or persons, bodies politique or corporate, their heires, successors, or assignes, as shalbe nominated or assigned by the said John Whitegift, his heires or assignes, according to such rules, statutes, and ordinaunces, as shalbe set forth, made, devised, or established, by the said John Whitegift, Archbishop of Canterburie, or his heires or assignes, in writinge under his or their or some of their hands and seales, not being repugnant or contrarie to the lawes and statutes of this realme: And furthermore, the said John Whitegift, Archbishop of Canterburie, doth, by theis presents, appoint, assigne, and lymitt, that all the profitts, commodities, and revenewes of all the landes, tenements, rents, hereditaments, and possessions, by theis presents mentioned to be geven, or hereafter to be geven, and likewise all the goodes and chattells hereafter to be geven to the relief of the same hospitall, shalbe converted, disposed, and employed, to and for the mayntenaunce and sustentation of the said hospitall, and of the said bodie and members thereof, according to the rules, statutes, and ordinaunces, hereafter to be lymited, assigned, or appointed, as is aforesaid: And further, the said John Whitegift, Archbishop of Canterburie, to and for the present endowment of the said hospitall, doth by theis presents (according to the purport and true meaning of the said acte of parliament) give, graunte, appoint, and confirme unto the said wardein and poore of the said hospitall of the Holy Trinitie, in Croydon, of the foundation of John Whitegift, Archbishop of Canterburie, and to their successors for ever, one annuetie, or yearly rent of tenne pounds by the yeare, of lawfull money of England, to be had and taken oute of all those landes and tenements of the said John Whitegift,

Archbishop of Canterburie, called or knowen by the severall names of Christian-field and Rye-crofte, conteyninge, by estimation, threescore and seventeen acres, scituate, lying, and being in the parish of Croydon, in the said countie of Surrey, to have and to receave the said annuetye or yearlie rent of tenne pounds, unto the said wardein and poore of the said hospitall of the Holy Trinitie, in Croydon, of the foundation of John Whitegift, Archbishop of Canterburie, and to their successors for ever, at the feast of St. Michael the Archangell and the Annunciation of the Virgin Mary, by even portions, yearlie to be paid: And if it happen the said yearlie rent of tenne poundes, or any parte therof, to be behind unpaid, in parte or in all, by the space of tenne dayes after any of the said feasts in which (as aforesaid) it ought to be paid; that then, and so often, it shalbe lawfull for the said wardein and poore of the said hospitall of the Holy Trinitie, in Croydon, of the foundation of John Whitegift, Archbishop of Canterburie, to their successors and assignes, to enter into the premisses, or any parte thereof, there to distrayne, and the distresse so taken to withhold, untill they be of the said rent and of the arrearages (if any be) fully satisfied, contented, and paid; which said landes and tenementes, called or knowen by the several names of Christian-field and Rye-crofte, are holden in soccage, and the said John Whitegift, Archbishop of Canterburie, is and standeth thereof seized of a good, sure, and lawfull estate in fee simple, absolutelie, to hym and his heires. In witnes wherof, the said John Whitegift, Archbishop of Canterburie, to this present deede hath putto his hand and seale.

Dated the five and twentith day of June, in the yeare of the reigne of our Soveraigne Ladie Elizabeth, by the grace of God, Queene of England, France, and Ireland, Defender of the Faith, &c., the one and fortithe.

JO. CANTUAR.

Sealed and delivered in presence of us,

RIC. LONDON.	JOHN PARKER.
JO. BATH. & WELLEN.	ED. BARKER.
J. W. FAWKES.	GEORGE PAULE.
JON. BOYS.	ABRAHAM HARTWELL.
JH. HYLWORTH.	RICHARD MASSINGER.
W. BENLEW.	FFOWLKE BOWGHTON.
FFRANCISCO BUTLER.	WILLM. BEESTON.
CHR. WORMEALL.	WILLM. SEGAR, *Norroy King of Armes.*
JOHN GILPIN.	
AYMAS NEVILE.	

Irr. in dors. claus. cancellar. infrascr. dne regine vicesimo sexto die Junii anno infrascript.

Per JOHEM SNELLING.

No. X.

Statutes, Constitutions, and Ordinaunces, devised by me, John Whitegift, Archbishop of Canterbury, Founder of the Hospytall of the Holye Trinity, in Croydon, in the County of Surrye, and given unto them of the sayde Hospytall, for the Order, Governemente, and Direction, touchinge the Lands and Tenements of the said Hospitall and all the Members thereof. (From a MS. in the Library at Lambeth, No. 275).

CAP. I.—*Of the Number of those that are to be maynteyned in or by the Hospitall.*

FIRST, I do ordeyne, that the number of the bretheren and sisteres of the sayde Hospitall shalbe ever thirtie at the least, and so many more, under xl in all, as the revenues of the sayde Hospitall, accordinge to the proportion in theis ordinaunces lymitted, may beare, untill all the severall roomes therein appointed for lodginge be replenished; of the which number of bretheren, one shall teache a common schoole in Croydon, in the schoole-house there by me buylded, and performe suche other duties as is appointed unto him in these ordinaunces and statuts: Provided always, that the yerely sume of tenn pounds owte of the revenewes be yerely reservid for reparations, sutes in lawe, and other necessary charges, &c., over and above the proportions hereafter lymmited. Item, yf any of the places of the poore brethren or sisters aforesayde, upon any occasion, shall happen to be voyde by the space of one hale monethe or more, or the place of the schoole-master by the space of three moneths or more, or yf yt shall please God so to blesse the hospitall, as that any other overpluss of revenewe shall remayne, when all the roomes as aforesayde shalbe replenished; the allowance that shalbe due unto suche voyde places, and suche overpluss, shalbe layde up and reserved safe in the common chest of the sayde hospitall, as a stocke to be imployed for repayringe, reedyfyinge, defence in lawe, or for other common charges.

Cap. II.—*That Women may be placed in the Hospitall.*

I ordeine, that the sayde hospitall may have women placed in yt, aswell as men; they the sayde women being poore, and qualifiede in lyke manner, forme, and degree, as is hereafter expressed in the statutes touchinge the seconde and third degree of those who are to be placed: Provided nevertheless, that at no tyme above one half parte of the whole number (not accountinge in this behalf the wardein nor the schoolemaister) shall consiste of women only: Provided also, that the poore wydowes of longeste contynewance in Croydon and Lambethe, beinge quallyfiede accordinge to the ordinaunce, shalbe prefferred before all others.

Cap. III.—*Who shall not be lodged or enterteyned in the Hospitall.*

No man nor woman shall lodge in the sayde hospitall, eyther wyfe, children, or others, not beinge members of the sayde hospitall; neither shall enterteyne any manner of person in the towne of Croydon, not being borne in the sayde towne, or there remayninge three yeres next before, whereby the towne may presently or afterwards be burdened, uppon payne of loss of their place in the sayde hospitall forever *ipso facto*.

Cap. IV.—*Of the Ellection and Placinge of the Members of the Hospitall.*

I doe ordeine, that, within one monethe (yf conveniently yt may bee) after yt shalbe notifiede by the warden, or otherwyse, that the place of the schoolemaster, or of any other of the poore brethren or systers of the sayde hospitall, is any waye become voyde, the Archbishop of Canterbury (for the tyme beinge) or, the see beinge voyde, the parson of Lambethe, and vicar or curate of Croydon, shall nomynate and place some one qualified accordinge to these ordinaunces, under his or theyr hande and seale; who, uppon suche nomination or placinge, shall wythoute delaye be sworne and admitted as is hereafter expressed: But if the parson of Lambethe and the vicar or curate of Croydon for the tyme beinge, when the see of Canterburye is voyde, shall not agree of some one quallyfyed accordinge to these ordinaunces, within one monethe after yt come to theyr knowledge that a place is voyde, then I will that the archdecon of Canterburye for the tyme beinge shall supplye theyr defecte therin, so that he name suche an one as is quallifiede accordinge to theis statutes.

Item, I ordeine and appointe, that the poore brother appointed to be the schoolmaster shall be a parson well qualyfyede for that function, that is to saye, an honest man, learnede in the Greeke and Lattin tongues, a good versifiere in bothe the foresayde languages, and able to wryte well (yf possible yt may bee); which poore brother appointed to that offyce, and quallified and placed as afore, shall have for his lodginge and dwellinge place, during the tyme that he contineweth schoolmaster, that howse which I have buylded for that purpose, adioyninge to the sayde hospitall, and nere unto the sayde schoolehouse, together with suche backsides and grounds as I have appointed to be annexed to the sayd howse, and which the present schoolemaster nowe enioyethe, and shall also have the some of twentye pounds yearely for his stipande, to bee payde quarterlye, together with other further comodyties of corne or wood, as hereafter shal happen to be allotted to other of the poore brethren of the sayde hospitall: I doe lykewyse ordeyne and appointe, that the howse which I have builded for the sayde schoolehowse, and also the howse which I have buylded for the schoolemaster, shalbe for ever imployede to that use onlye, and to no other.

The wardein frome tyme to tyme, so often as the place shalbe voyde, shalbe one of the poore brethren of the sayde hospitall, and shalbe appointed by me the founder, during my lyfe; and after my deathe, and the deathe of suche wardeine as is allreadye appointed by myself, suche one of the poore brethrene shalbe chosen after moreninge prayere, in the chappell of the sayde hospitall, within seaven days after every suche vacation, as the greater parte of all, that is to saye, of the schoolemaster (yf that place be then full) and of the other poore brethrene recconed together, and then present, shall chose to be wardein: But yf the voyces happen to be even, then suche poore brother shalbe wardein, with whome the schoolemaster, as is aforesayde, did give his voyce: But yff the greater parte of the hole companye cannot, within the tyme aforesayde, agree upone one, then suche brother shalbe wardeine as yt shall please the Archbyshopp of Canterburye, or (the see beinge voyde) the vycar or curate of Croydon, to nominate, uppon advertysemente by lettres thereof, to be gyven by the senior poore brother, or schoolemaster, yf any suche bee.

The office of the wardeine shalbe, to keepe one of the keyes of the comon chests and dore of the evidence-howse; to procure that the gates be locked and opened at due tymes apointed; and that the keyes, on nyghtes, be broughte unto him; to be present at all admissions and payinge of wages; to see that all enteryes be duly made in the lidger booke, and the evidence well and safelye layde up and kepte; to keepe the keyes of the voyde lodgings, and to deliver them to the next brother or syster newlye appointed; to looke in tyme to reparations, and to all other good husbandry of the hospitall; to foresee that fyre or candells be not daungerouslye kepte; to require and exacte of eache one of the poore brethren and systers the observation of the ordinaunces and statuts; and suche as be necligente and faultye gentlye to admonishe them, or, yf the qualitie of the faulte so require, to complayne of the delinquents unto the Archbyshopp of Canterburye, or, the see archiepiscopall beinge voyde, unto the Custos Spiritualitatis of the see of Canterbury for the tyme beinge, to whome I give authoritie to redress the same, accordinge to his dyscretion.

CAP. V.—*Who shalbe chosen into the Hospitall, and the severall Degrees of them.*

In the firste degree of the poore brethren that are to be placed in the sayde hospitall, I ordeine, that suche men, beinge honest, well reported, aged lx yeres at the leaste, poore, and not otherwyse able to get their livinge, who have served in howsholde the Archbyshopp of Canterbury, shalbe preferred before all others, so that there be not above three of them of the sayde hospitall at one tyme; and before all others, those that have served myself, or be alm unto me, being impotente, and not able otherwyse to gett theyr livinge, thoe they be under the age before mentioned, or above the foresayde number.

I ordeine in the seconde degree, that, before others, suche men and women of the parishes of Croydon and Lambethe shalbe preferred, beinge honest persons and of goode reporte, of the age of lx yeares at the leaste, and of the pooreste sorte, being impotente, and not otherwyse able to get theyr livinges.

I doe ordeine, that the thirde degree, in defecte of all the former, before all others, shalbe preferred thither such poore honest persons of good reporte, beinge of the age of lx yers at the leaste, as be of suche parishes within the countye of Kent, whereof the parsonage is appropriate unto the archiepiscopall see of Canterbury; and before others, those cheyfly of suche parishes whereof the sayde see dothe receave most revenewe: Provided, that this ordinaunce be not extended to any suche poore as are otherwyse provided for in any of the sayd parishes, or in any other places. Neither shall this ordinaunce, or any thinge conteyned in this chapter, be extended to that poore brother who is to be appointed for the scholemaster.

Cap. VI.—*Off the Admittance of the Members of the sayde Hospitall.*

The sayde schoolemaster, warden, and every other poore brother and syster of the sayde hospitall, not being dumb or deaffe, shall, in the hall of the hospitall, before theyr placinge, in the presence of the schoolmaster, wardeine, and senior brother, or any two of them, whereof the warden to be one, yf that place be then full, not onlye take a corporall oathe of obedience and allegiance to the Queens Majestie, her lawful heyers and successors, so as by lawe is prescribed; but also this oathe hereinfollowinge, viz.—

> "I, A. B., from hensforthe, as longe as I shall remayne a member of this hospytall, shall and will, by God's assistance, do my beste endevor to obey, performe, and fullfill the ordinaunces and constitutions of the same, insomuche thereof as dothe concerne me: I shall not any tyme hereafter wyllingly procure or give assente unto any endaungeringe, hurte, or endammaginge of the sayde hospitall, eyther in the state or any the hereditaments, or in any the moveable goods thereof; but, to my best powre and skyll, shall defende and sett forward the good estate, commodity, and wellfare thereof, whilest I live: So help me God in Christe Jesus."

CAP. VII.—*Off the Office of the poore Brother that is appointed to be Schoolemaster.*

The schoolemaster shall freelye teache suche of the chilldren of the parishe of Croydon, withoute exactinge any thinge for theyre teachinge, as are of the poorer sorte, suche as shalbe so accounted by the vycar or curate of Croydon, and two of the better sorte of the inhabytants in Croydon: But yet yt it shalbe lawfull to and for the sayde schoolemaster to receave that which is voluntaryly bestowde uppon him by any of the sayde poorer sorte of parishioners, and for the children of suche as be of the better sorte of the parishioners of Croydon: Yf the sayde schoolemaster shall exacte tomuche for theyr teachinge, or refuse to teache them, the same shall be ordered or moderated by the Archbyshopp of Canterburye for the tyme beinge.

Bothe the saide schoolemaster and scollers shall frome tyme to tyme be ordered, governed, and directed, by suche prescriptions and ordinaunces, in all points, as by me the founder of the sayde hospitall shalbe in my lyfe-tyme devised, and after by my successors Archbyshopps of Canterbury, soe that allways the saide ordinaunces of my successors be not contrarye to my ordinaunces.

CAP. VIII.—*Of the Yearelye Proportion of Allowance of the Members of the Hospitall.*

Firste, The custos or wardeine shall have yerely six pounds extraordinary allowance; the schoolmaster, beinge a member of the sayde hospitall, shall have yerely twenty poundes; and every other poore brother and systere of the sayde hospitall shall have yerely five pounds apece allowance, over and besyde such wood, corne, and other provisions to eache of the brethren and systers, as nowe or hereafter shall, by God's providence, and by the devotion of charitable-minded men, be allotted unto the sayde hospitall; the expence in monye to be payde unto them every quarter, and the yere to begin at the feaste of St. Mychaell tharchangell.

Cap. IX.—*Off the Bookes and Register of the Hospitall, and of Receipte of Rentes, &c.*

Allso, there shall be a fayre lidger booke made and kepte in a chest, with locke and keye, standinge in suche a chamber in one of the gatehowses as I shall appointe, wherein by the schoolemaster shalbe entered and regestered the names, ages, qualietye, and tymes of every admittaunce of warden, poore brother and syster, and the tymes of theyr deathes or removeinges; also, there shalbe another fayre lidger booke in the sayde chest, wherein shalbe fayre entered the coppyes of all leases and other graunts that be alreadie in lease, or hereafter shalbe made by the sayde hospitall; and a third lydger booke in the sayde cheste, wherein shalbe fayre enterid the names and severall gyfts of all their benefactors, the inventorye of all theyr moveables; and generally, all other things of momente, from tyme to tyme, shalbe regestred, that do any way concerne the sayde hospytall, as in theis ordinaunces hereafter is expressed.

The sayde custos or wardeine, and the schoolemaster, and the claviger or chest-keeper, or two of them, shall receave the rents of the sayde hospitall, and dystribute the same to eache, accordinge as is afore lymmyted.

CAP. X.—*Off daylie Prayer to be used in the Hospitall, and other Exercises of Piety.*

The schoolemaster, besides teachinge of his schollers, and makinge enteryes into the lidger bookes, and doinge othere duties appointed unto him in theis ordinaunces, shall saye publyke prayers, moreninge and eavenninge, in the chappell of the hospitall, to the brothers and systers, on all dayes (beinge workinge days) excepte Wednesdays and Frydays in the forenoone, and Satterdays in the afternoone.

All the bretheren and systers of the hospitall, beinge at home, not sicke, nor otherwyse lame, and unable to go so far, and excepte the porter, and some one in course to tarry att home to keepe the howse in abcence of the reste, shall, on the Saboth days, Feastivall days, Wednesdays and Frydays at morninge and eaveninge prayers, and upon Satterdays at eaveninge prayre, resorte orderlye by two and two togethar to the parishe churche of Croydon, there to pray devoutlye with the reste of the congregation, and namelie for the preservation of the Queenes Maiestie and her realmes; to give God thanks for theyr founder, benefactors, and for all other God's benefytts, and to here the Worde of God; and there to be partakers of the Holy Sacramente of the Lord's Supper, at the leaste thrise everye yere: Provided, that this ordinaunce be not extended to the scoolemaster, for Wednesdays and Frydays, nor for the manner of goinge to the churche.

Allso, I doe ordeine, that, on the reste of all the workinge days, moreninge and eaveninge prayer shalbe accordinge to the Book of Common Prayre, to be sayde by the schoolemaster in the chappell of the hospitall, unto whyche all the bretheren and systers shall duly resorte, unless they be sicke, or otherwyse so lame as that thereby they shalbe unable to come thyther; and that some one of the companye be weekelye appointed by the wardein to note suche as are absente frome prayers, and to gyve theyre names daylye to the sayde warden; the sayde moreninge and eaveninge prayers to be sayde att suche hours as the sayde schoolemaster shall thinke fytteste. Yff any, otherwyse than as afore, shall absente themselves from prayers, eyther att the parrishe churche or at the chappell in the hospitall, wythoute a sufficiente cause, to be allowed by the wardeine; for the firste tyme, suche shall forfeite one ob; for the seconde, id.; and soe for everye tyme after, in eache monethe as is aforesayde, to be abated and defaulted frome theyr allowaunce at the paye daye happeninge next after suche theire defaulte, one thirde parte of suche forfeytures to be imployede unto the porter for that monethe, and the other two thirds to the common cheste: But in case any one, withoute such cause as is aforesayde, shalbe founde to have forfeited, in eache of any fowre monthes happeninge in one yere, the sayde yeare to beginne to be accounted at the feaste of Saincte Mychaell, so muche as shall amounte above a thirde parte of his monethlye allowaunce; then, for every suche default happened in any one yeare, the necligente person shall have one solemne admonition given him by the wardeine and schoolmaster, which shalbe entred in the lidger booke; but yf, after thre admonitions, the same partye shalbe founde agayne to have offended in the same kinde, he or she shall then, for a necligente and an incorrigible parson, be expelled frome the hospitall, never to be receavide theare agayne.

CAP. XI.—*Of the Porter, and his Office.*

I doe ordeine, that the wardeine shall, uppon the firste daye of every monethe, nominate one of the brethren, whome he shall think fittest, to be porter for the reste of that instant moneth.

The office off the porter shalbe to ringe a lytle bell twice eache moreninge and eaveninge (unto prayers), the one ringinge to be a quarter of an houre afore the other; allso, to receave the keyes of the gates of the hospitall frome the wardeine eache moreninge, betwixte the feaste of Thannunciation and Sainct Mychaell, aboute seaven of the clocke, and then to open the foregate, and to shott yt at eighte of the clok in the eaveninge during that tyme, and to carry the keyes to the wardeine eache nighte; and from the feaste of St. Mychaell unto the feaste of Thannunciation, toe open the gates aboute eighte in the moreninge, and to shutt them at seaven of the clocke at nighte, and then to carry the keys to the wardeine.

Cap. XII.—*In whate Wordlye Busines they of the Hospytall may exercise themselves.*

It shalbe lawfull for any of the brothers and systers, havinge skill in any manuall trade, to worke on the same, within the hospitall or withoute, thereby to get some parte of theire lyvinge; or for any of the brothers and systers, beinge able in bodye, to exercise themselves in any honest handy labor of the bodye abrode; yet so as that, without especiall leave of the wardeine, they do not in suche respecte lodge owte of the hospitall above one nighte in any one weeke. Provided always, that none of them kepe any alehowses or vitalinge howses, or suche lyke, eyther within the sayde hospytall or without, uppon payne of losinge theyr places *ipso facto.*

It shall not be lawfull for any member of my sayde hospitall, eyther by themselves or by any other, to begg or crave of any parson or parsons, eyther wythin the towne or ellswhere. Suche as shalbe fownde so to doe, after two admonitions given by the wardeine, who by vertue of his oathe (after notice had thereof) shall charge him to performe, to bee foorthwith expelled the sayde hospitall. Nevertheles, yt shalbe lawfull for them to receave the almes and benevolence of any parson or parsons voluntarelye offeringe the same, wythoute such kinde of begginge or craving; the same to be distributed in common to the poore of the saide hospitall, when yt shall come to such a quantitie as the reste may be partakers thereof. In the meane tyme, the same to be put in a box prepared for that purpose; of which box there must be two severall locks and keyes, the one keye to be kepte by the wardeine, the other by the schoolmaster for the tyme beinge; and once in every quarter the box to be opened in the presence of most parte of the brethren and sisters, and the mony to be devyded amonge the hole companye: Provided allways, that the schoolemaster shall have noe parte of this almes and benevolence: Provided likewyse, that yf any thinge be bestowede uppon any perticuler parson, in respecte of kindred, sickness, or other impotencie, that wholye shall goe to the partie on whome yt is perticulerlye bestowed.

Cap. XIII.—*What Crimes and other Inconveniences are to be avoyded, and uppon whate Pennalties.*

If anie brother or syster shalbe convinted of any kinde of incontinencie, forgerie, periurye, obstinate in heresye, sorcerye, or of any kinde of charmmynge or witchcrafte, or of any cryme by the lawes punisheable by loss of lyefe or lyme, or of eare, or shalbe publiquely sett on the pillorie, or whipped, for any offence by them committed; or shall obstinately refuse to frequente devine service by lawe established; immeadiatelye thereuppon, and uppon confession, or convinction, suche brother or syster by the Archbyshopp of Canterbury, or by some to be deputed by him, or, that see beinge voyde, by the parson of Lambethe, and vicar or curate of Croydon, shalbe displeaced, and shall never be receavid in thether againe: Yff anie brother or syster shalbe a blasphemer of God's Holy Name, a swere, a gamester at any unlawfull game, a drunkard, or an hauter of taverns or alehowses, a brawler, fighter, contentious parson, scolde, or sower of discorde, and thereof shalbe convicted, by confession or honeste proofe, before the vycar or curate of Croydon, the wardeine of the sayd hospitall, the schoolemaster, or any two of them; suche offender shall, for the fyrste tyme, have a solemne admonition given, to be entred in the lidger booke; for the seconde tyme, shall forfeyte one moneth's allowaunce to the common cheste of the hospitall, and shall have another solemne admonition given as before; and yf he or shee offende in the lyke the third tyme, or in any other, the offences here named, to be expelled the hospitall forever.

Every brother or syster, withoute sufficient cawse, to be allowede by the wardein, shall nightly lodge within the hospitall, uppon payne to forfeyte owte of the next monethes allowaunce, for the first tyme anie one yere, two pence; for the seconde offence in the same yere, fowre pence; for the thirde suche offence, eighte pence; for the fowrthe suche, two shillings; for the fifte suche, the hole next monethes allowaunce; all which forfeitures shalbe appliede to the common cheste; for the sixt suche, to have a solemne admonition given, to be entered into the lydger booke; for the seaventhe suche offence in any one yere, which shalbe accounted to begynn at the feaste of Saincte Mychaell, another admonition lykewyse to be given and entered; and for the next suche faulte happeninge within the space of the same yere, to loose his or her place in the hospitall: Savinge that yt shalbe lawfull for any brother or syster, havinge a good cawse, and with lycens of the wardeine, the sayde licens and the day of theyr goinge forthe beinge first entered in the lidger booke, to be away for the space of two moneths in any one yere, beginninge att Mychalmas, eyther alltogether or at severall tymes; and savinge, that the schoolemaster shall not be any waye comprysed in this ordinaunce.

CAP. XIV.—*Off Care to bee had of suche as be Sicke or Impotente.*

For tendinge and comfortinge the sicke, in tyme of theyr sickness, and those that be impotente through age, or otherwyse be unable to help themselves, I do ordeine and appointe, that not onely all the systers shall, from tyme to tyme, do their carefull endevours towards them, even as themselves wolde wyshe to be respected by others in their owne extremities; but also I do ordeyne, that two of the systers, whome the wardeine shall thinke most fitt for that purpose, and shall nominate yerely the next day after the feaste of St. Mychaell, shall not refuse to looke more perticulerlye and especially to that Christian dutie as to theire owne more peculier charge and office, who shalbe called the Relievers of the Impotente, and, havinge well and carefully performed that charge, shall have, at the ende of eache yeere, six shillings and eighte pence a yere, in augmentation of their allowaunce, owte of the common cheste; and yf any so appointed by the wardeine shall refuse to take that charge uppon them, then the paretie so refusinge to be debarred frome receavinge any waygers or allowances for that yeare, the same to be put in the common cheste, to the use of the hospitall.

CAP. XV.—*Off the Howse of the Evidences, Chestes, and Comon Seale.*

Whereas I have allotted owte a speciall roome in the gatehowse next unto the streete, for keepinge of the evidences of the lands and revenewes of my sayde hospitall, and for other thinges of some momente, beinge not of dayly use; I doe ordeine, that in the sayde roome shalbe one cheste wythe three lockes and keyes of severall wardes and fashions; one keye whereof to be kepte by the wardeine, another by the sayde schoolemaster, and the third by the auncienteste brother, soe he be able to goe and walke abroade, or ells the next in auncientye that is able; in whiche cheste shalbe kepte the comon seale, one coppy of theis ordinaunces, and suche stocke of mony as, yearlye remayninge after all allowaunces, shalbe reservyd for reparations and for other necessarye disbursmentes.

I doe ordeine also, that in the same roome there shalbe one other cheste, wherein shalbe kepte the foundation and donation of the hospitall, and all other evidens whatsoever, well sorted, accordinge to the severall percells of landes, into severall greate boxes, superscribed wythe papers of direction; and also in the same cheste shalbe put all rentalls, surveys, terrars, with buttalls and roundes, courte deedes, yearleye accounts of the hospitall, and counterpartes of leases. This cheste shall have three lockes, and three keyes of several wardes and fashions; one keye to be kepte by the wardeine, another keye by the scoolemaster for the tyme beinge, and the thirde keye by one of the bretheren, to be yearlye chosen by the more parte of the bretherene of the sayde hospitall.

I ordeine, that noe parcell of evidence shall at any tyme be taken forthe thence, but upon especiall occasion; and then also not to be longer kepte frome thence then necessary occasion for the use thereof shall require.

There shall also remayne in the sayde cheste a paper booke, wherein shalbe entered the parcell of all evidences from tyme to tyme taken fourethe, the day and yere when, and to whose hands yt is to be delivered, and for howe longe tyme as is presupposed; and the day also and yeare shalbe entred, when and by whome suche parcell of evidence is redelivered in againe.

Cap. XVI.—*Howe there Lande, &c., shalbe demised, and with whate Governaunce, &c.; howe theire Woodes are to be kepte, and bothe Landes and Woodes surveyede.*

No lease nor other graunte shalbe made of any landes, tenements, or hereditaments belonginge to the sayde hospitall, unless bothe the wardeine, the schoolemaster, and also the greater parte besydes them of the reste of the poore brethren, shall yelde their consents thereunto; nor unless the accustomed yerely rente thereof (or more) be thereuppon reservid, and payable quarterlye, or at the leaste halfe-yerely, att or within the sayde hospitall; nor yf suche lease or graunte be above one and twentye yeres frome the makinge of the same, and wyth reservation of all timber-trees; nor unless in the sayde lease be conteyned trewe and perfecte percells and quantitye of landes, by common estimation, with the buttalles and boundalls thereof (yf convenientlye yt may be); also, in every suche lease or graunte shall a provisoe be conteyned, that the farmer shall paye the rente at the hospitall, wythin twentye days next ensuinge any one rente-day lymeted for payinge thereof, wythoute any demande to be made: Furthermore, in eache lease or graunte to be made, shall covenaunts to the effecte followinge be conteyned; that is to saye, firste, that the leassee, at his owne proper costs and charges, not only repayr, and, yf neede be, redifie all edifices thereuppon; and so well redifiede and repayred shall leave them at the ende of the terme; but also shall, frome tyme to tyme, hedge, fence, dyche, and scowre, accordinge to the usuall course of husbandry of the countrye where the sayde lands shall lye: Secondlye, that the leassee shall beare, pay, and discharge, or save harmles, the sayde hospitall, of and frome all charges, ordinarye and extraordinary, goinge oute or to be payde by reason of the landes demysed, or any parte thereof: Lastlie, that the leassee, betwixte every eighte and nynethe yeares of the sayde terme, shall make or cause to be made, and wrytten fayre in parchmente and deliver upp to the wardeine at the hospitall a trewe and perfecte terrar, conteyninge the name and quantitie, by estimation, of every percell of ground demysed, the names of the scituation and lyinge of the same towards other lands, and the names of the presente owners and terre tenaunts of the lands which are of any side abuttinge uppon the grounds demysed. Also, I doe ordeyne and appointe, that the sayde hospitall, uppon any reservation or otherwise, shall not encrease the rents or revenues of those lands I leave, or shall give them any higher or greater proportion then as the rents thereof now are, and accordinge to that rate they are nowe lett for.

Allso, I doe ordeine, that, in renewinge and lettinge of leaseas, the presente farmers be allways preferred, doinge reasonably for the benefite of the hospitall, as other men will doe; and amonge the reste, I wyll have those especially favoured who have theire leases from myself.

And allso, I do ordeine and appointe, that such mony as they shall rayse or make in fines uppon leases, or uppon sale of woods or trees, or by overpluss of theyr yearelye reavenues or otherwise (all necessary charges being deducted) shalbe layde upp in theyr comon treasorie, and kepte together untill yt wyle or shall amounte to the sume of a hundrethe pounds; and then the overpluss of that sume of a hundred pounds shalbe equalye devided by the wardeine and schoolemaster for the tyme beinge (calling to them two of the senior brethren) amongste all the poore brethren and systers of the sayde hospitall, and then to have theyre equall portions wyth the reste; which sayde some of one hundrethe pounds or under shalbe preserved and kepte in the place aforesayde, for any extraordinary occasions, as, for sutes in lawe, reparations of the sayde hospitall and schoolehowse, and suche lyke; and as the same shalbe by such charges demyneshed, so to be allways replenished wyth lyke receiptes, as they shall come in or be receavide.

Provided alwaies, and my meaninge is, that for all woodes belonginge to the sayde hospitall as shalbe lefte unleassed, that the yerely value thereof shalbe taken as parte of theyre yerelye revenue, and not to be raysed in stocke, as in the laste article, but onely such trees and woods as are not annuall in profitt, but maye be comodiouse in tyme.

Cap. XVII.—*The Revennues off the Hospitall, by whome to be received and disbursed, and of a yearely Account.*

All the rentes and revennues shalbe payde in the hospitall, to the handes of the wardeine and schoolemaster, and the other claviger, who all shall write an acquitance for eache receipte. But, yf eyther of theis places be voyde, or eyther of them be so sicke, or otherwyse absente, that they cannot be presente at suche paymente, then shall yt be done in the presence of the next two poore brethren in auntyentry that are able to stirr abroade, they callinge unto them (yf neither of them cann wryte) some of the brethren that cann write; and in defecte hereof, some other honeste person who is able to wryte; and ymediatlye upon suche receipte, an entrye thereof shalbe made into the lidger booke; and then shall the mony be presentlye layde up in the common chest, there to remayne tell ther be occasion of disbursmente thereof againe: In the afternone of the firste day of every quarter, the thre clavigers, or cheste keepers, taking forethe of the comon cheste so muche mony onlye as then is to bee disbursed, shall presently, in the hall of the sayde hospitall, paye unto every one of the brethren and systers, or, in case any be sicke or owte of towne, to theyr attorney, beinge one of the brethren, theire severall due allowances, makinge presently a note of the recept thereof in the lidger booke.

When any other occasion besydes the quarterly wages dothe happen for disbursmente of monye, as, for reparations, sutes in lawe, or suche lyke, the same shall also, wythe the day, occasion, and in whose presence and to whose hande yt was delivered, together wythe the hande or marke that receivid yt, be entred in the lydger booke.

Everie yeare, on the fourethe day of December, the schoolemaster, in the presence of the other two clavigers, and of all the other brethren and systers that cann and wilbe presente, havinge caste up afore and sumed all accounts, aswell of receptes as of disbursments, for the yere endinge at the feaste of St. Mychaell next afore, shall declare unto them and goe over the perticulers of all the accounts for the sayde hole yere, that the estate of the hospitall howe that yt standethe may yerely so appere unto every one of them; and yf any arrerages be then fownde to be in any the accomptants hands, the same shall, eyther presentely, or within thre days at the furtheste, be delivered in unto the clavigers, and shalbe layde upp in the common cheste, uppon payne of loss of the next monethes allowance unto the comon cheste of him that shalbe so behinde and in arrerages; but, yf within twentye eighte days next after the sayde three days expired, the whole arreragis shal not be payde, then suche one shall loose his place *ipso facto*, and be suede in lawe for the arrerages remayninge in his hands.

CAP. XVIII.—*Off the Reparations of the Hospitall, and by whome and when to be performed.*

Iff anie glasse windowe be broken, or other decaye, by wyllfullness or ne̜cligence, be made in any private roome of the hospitall, the same, uppon wareninge given by the wardeine, shalbe amendid within one monethe, by him or her, and at his or her charges whome the roome is, uppon payne to loose foure pence for every weeke after tell yt be mendid; yff the glass of any publique roome be broken, and not beinge knowne by whose default yt was done, yt shalbe repayred againe by the overseer of the work, at the publik charges of the hospitall; the howse allotted for the schoolemaster to dwell in shalbe repayred at the costs and charges of the schoolemaster, uppon suche penalties as the Archbyshopp of Canterbury for the tyme beinge shall thinke conveniente.

Everie yeare, the nexte day after the feaste of St. Mychaell, the wardeine and the schoolemaster shall appointe one of the brethren (thoughte to be moste fitt thereunto) to be overseer of the workes and reparations of the hospitall and schoole-howse, for the yere ensuinge; whose offyce shalbe, dyligente to provyde that noe decays be left unrepaired, but amended, betwixt the feaste of the annunciation of the Blessed Virgin and of St. Mychaell tharchangell; but, yf any tyle be fallen off, or suche decaye happen as cann abyde noe delaye, the same to be wythe all convenient speede repayrede, thoughe yt be in the winter tyme.

Cap. XIX.—*Howe the Wardeine and Schoolemaster shalbe censured, yf he or they shalbe fownde to be necligente in performynge suche Duties as by theis Ordinaunces are imposed uppon him or them.*

I doe ordeine, that yf the wardeine of my sayde hospitall or schoolemaster shalbe founde to be necligente in performyng the charge by these ordinaunces imposed uppon him or them; then, uppon notyce thereof given to the Archbyshop of Canterburye for the tyme beinge, suche ponishmente shall be inflicted upon him or them, as the sayde Archbishop, in his dyscretion, shall thinke conveniente.

Cap. XX.—*Touchinge the Chambers which the Fownder reservethe to himself.*

Item, I doe ordeine, and my will is, that the chamber over the hall, and the two chambers over the inner gatehowse, shalbe reservid to myselfe, and to my owne use, during my liefe; and after my deathe, my will and meaninge is, that my executors shall have and enioy the sayde chambers for one hole yere next after my deathe; and that, after the experation of the same yeare, my brother George Whitegifte shall have and enioye the same chambers duringe his lyfe: Provided allwayes, that he do not assigne the same over to any other, nor place any therein, unless yt be some of the members of that my hospitall; and after his deathe or relinquishmente, the same chambers to remayne to the wardeine of the sayde hospitall, and his successors for ever.

Cap. XXI.—*Off the Founder, Visitor, and Cheiffe Governor of the Hospitall.*

It shalbe lawfull for me, the nowe Archbishop of Canterbury, founder of the sayde hospitall, to abrogate, add unto, chaunge, or alter, theis ordinaunces, and to place or displace anie member thereof, wythe cause, or wythoute cause to be rendred thereof unto any other, to dispose of the lodgeings in and wythoute the sayde hospitall, to lett leases, and helpe to governe the same, according as shall please me, duringe my natural lyffe, withoute any other persons intermedlinge therein. After the death of me the sayde fownder, then the Archbishop of Canterbury for the tyme beinge, bye himself or other whome he shall appointe, shall have full powre and authoritie, from tyme to tyme, not onely to interprett any doughte arisinge out of the ordinaunces which bye me the founder shalbe lefte unto the hospitall, but allso shall have full powre, libertye, and authoritie, to ponishe, confine, and remove anie member thereof convicted accordinge to theise ordinaunces.

I doe ordeine, that my successors, Archbishops of Canterbury, shalbe the continewall patrons, governors, and visitors of the sayde hospitall; earnestlye requestinge them (in the bowelles of Christe) to have, frome tyme to tyme, a fatherly and compassionate care of theire good estate, and of the poore members thereof; and that they wolde be pleased from tyme to tyme (as occasion shalbe offered) to compose theyr controversies, to protecte, advise, order, governe, and direct them, and, when neede shall require, by themselves, or bye such discreite persons as they shall thinke fitt, in personn freelye to visite the sayde hospitall, and to enquire bothe of the publique state of itt, and also of the private demeanure of every perticuler member thereof, by suche a course as the lawes dothe allowe; which visitation I wolde hartilye wishe might at the leaste every third yeare be performed, whether there seeme anie necessarie occasion thereof or noe.

Also, I doe ordeine, that once in the yeare at the leaste, within tenn days after the feaste of St. Mychaell, theis ordinaunces and statutes shalbe openlye reade in the chappell of the sayde hospitall, and all the bretheren and systers admonished to be theare presente. And, for the better governemente of this my hospitall, becawse I understande of some discordes breedinge amongeste the poore that are therein allreaddy by me placed, for wante of some discretion and understandinge to directe them in observinge the orders and statutes of this my hospitall; therefore, I doe ordeine and appointe, that the vicar of Croydon allwayes, for the tyme beinge, shall have the oversighte of the wardeine and poore there, aswell to directe them in the observinge, as to ponishe them accordinge to the sayde lawes and statutes of my hospitall (yf they) in theyr severall places and offices do not theyr duties accordinglye; and to this purpose, I will allwayes have one coppie of theis my sayde ordinaunces and statutes to remayne wythe the sayde vicar for the tyme beinge, harteley prayinge him, and in the name of God charginge him, duly to performe the truste by me reposed in him: Provided allways, that this statute and ordinaunce doe not derogate any authoritie from the Archbishop of Canterburye for the tyme beinge, given unto him by my former statutes and ordinaunces, or due unto him as vysytor of my sayde hospitall.

JO. CANTUAR'.

And whereas Samuell Fynche, the nowe minister of Croydon, hathe taken verye greate care and paynes about the buyldinge and erectinge of this my hospitall; and in hope that he will continewe the lyke care of the same after my deceass, and endevor the best he maye to see my statutes and ordinaunces kepte and performed, accordinge as by my late letters I have authorised him;

I doe therefore nowe ordeine and appointe, that he shoulde have, during his life, yearely payde unto him, after my deceass, the some of six pounds thirtene shillings and foure pence, quarterly, at suche tymes as the other pensions beforementioned to the poore are to be payde att. And after his deceass, the sayde pention to ceasse, and not to be payde to any other; but to remayne to the use of my said hospitall only.

JO. CANTUARIEN.

No. XI.

A Case resolved touching my Hospitall. (*From a MS. in the Lambeth Library, No.* 275).

THE Archebishopp of Canterburie houldeth his mannour of Croydon of the Queene in frankallmoyne.

One that houldeth of the same Archbishopp as of his said mannour in soccage geeveth his tenancy to the Bishopp and to his heires.

Quere, howe this tenancy nowe in the hands of the Bishopp is holden, in regarde of his severall capacities?

Whether alltogether suspended, or of the Queene *in capite*? and howe, upon the bargaine and sale, or other like conveyance, the tenor will then bee?

And quere, which is the surest meanes for the Bishopp to give his tenancy to his Hospital, because the statute alloweth no *capite* lands to be given?

It is holden verie clere, and without any doubte, that by the purchase of the tenancy there is no alteration of the service, but that the tenancy is holden in soccage, *ut prius*, of the seignory suspended; for that it is in the Lord Archebishopp in divers respects, and his estate in the tenancy is free, and in the seignory but for life in a manner, and that in this case there is no tenure of the Queen's Majestie during the suspension.

By the opinion of

The Lord ANDERSON, Chief Justice of the Common Pleas.

Justice GAWDIE, Secondarie Justice of the King's Benche.

Justice WAMESLEY, Secondarie Justice of the Common Pleas.

Justice GLANVILE, Puisne Justice of the Common Pleas.

Per me W. COMBE.

No. XII.

Negotium Dedicationis & Assignationis novæ Capellæ sive Oratorii Hospitalis vocati Hospitale Sanctæ Trinitatis, in Croydon, ex fundatione Johannis Whitegifte, Archiepiscopi Cantuarien. ad Usum Pauperum ejusdem Hospitalis noviter erecti & fundati. (From a MS. in the Lambeth Library, No. 275).

Die Lunæ, nono viz. die mensis Julii, anno Domini millesimo quingentesimo nonagesimo nono, dictus Reverendissimus in Xp̃o pater, fundator ejusdem Hospitalis & Capellæ sive Oratorii, ac Ordinarius illius loci, in palatio suo de Croydon, in præsentia mei Thomæ Redman Notarii publici specialiter assumpti, &c. commisit vices reverendis in Xp̃o patribus ac dominis dñis Richardo London' et Anthonio Cicestren' respective ep̃is, commissionis et deputationis ad dictam novam Capellam sive Oratorium hospitalis vocati Hospitalis Sanctæ Trinitatis, in Croydon, ex fundatione Johannis Whitegifte, Archiepiscopi Cantuariensis, divino cultui, et divinorum celebrationi, ac verbi Dei prædicationi, quantum de jure possit, et per leges et statuta hujus inclyti Regni Angliæ licebit, dedicandum et assignandum, ac ad nominandum eandem Capellam sive Oratorium per nomen Capellæ sive Oratorii Hospitalis Sanctæ Trinitatis, in Croydon, ex fundatione Johannis Whitegifte, Archiepiscopi Cantuarien'; necnon ad procedendum, decernendum, et faciendum in dicto negotio, juxta statuta, leges, canones, ordinationes, ritus, et consuetudines Ecclesiæ Anglicanæ in ea parte stabilit', et nunc usitat' et observat'. Et deinde dictus reverendissimus pater decrevit ut tempore dedicationis ejusdem Capellæ sive Oratorii divinæ preces celebrentur, ac sacra concio publice ibidem habeatur, et assignavit pro ea vice in concionatorem magistrum Thomam Monforde, sacræ theologiæ doctorem, ad concionandum et prædicandum verbum Dei in dicta Capella sive Oratorio prædicto, et constituit diem in quo præmissa peragentur, viz. diem decimum præsentis mensis, Anno Domini 1599 prædicto; præsentibus tunc ibidem, ac præmissa omnia videntibus et audientibus, atque ad eadem testificandum specialiter requisitis et rogatis, venerabilibus viris Will'mo Barlow*, sacræ theologiæ professore; Johanne Parker, armigero; Edwardo Aylworth, armigero; Will'mo Thornhill, artium magistro; Michaele Murgatrode, Francisco Butler, Will'mo Beeston, et Richardo Massinger, generosis, &c.

THOM. REDMAN, *Notarius Publicus.*

Quo quidem decimo die mensis præsentis Julii anno Domini 1599 prædicto adveniente, inter horas octavam et duodecimam ante meridiem ejusdem diei, præfatus reverendus pater dñs Richardus London' episcopus in dicta Capella sive Oratorio personaliter præsens et sedens, ob honorem et reverentiam dicto reverendissimo patri debitam, onus executionis dictæ commissionis sive deputationis in se acceptando, ac virtute ejusdem commissionis sive deputationis procedendo, dictam Capellam sive Oratorium divino cultui, divinorum cele-

* Afterwards Bishop of Lincoln.

brationi, ac verbi Dei concionandi et proponendi usui, quantum de jure potuit, et per statuta et leges hujus inclyti Regni Angliæ licet, dedicavit, per nomen Capellæ sive Oratorii Hospitalis Sanctæ Trinitatis, in Croydon, ex fundatione Johannis Whitegifte, Archiepiscopi Cantuariensis, et sic dedicatum et assignatum esse, et in futurum perpetuis temporibus remanere debere, palam et publice denunciavit; eamque Capellam sive Oratorium per nomen Capellæ sive Oratorii Hospitalis Sanctæ Trinitatis in Croydon, ex fundatione Johannis Whitegifte, Archiepiscopi Cantuariensis, perpetuis temporibus futuris nominandum et appellandum fore decrevit, et sic nominavit et appellavit. Quibus sic gestis, tunc et ibidem preces Deo Optimo Maximo juxta formam descriptam in libro publicarum precum authoritate parliamenti hujus inclyti Regni Angliæ stabilito celebratæ fuerunt, atque immediate verbi Dei concio per dictum magistrum Thomam Monforde sacræ theologiæ doctorem (ut præfatus) designatum publice facta fuit; præsentibus tunc et ibidem, ac præmissa omnia videntibus et audientibus, et ad eadem testificanda specialiter requisitis, reverendo patre d̃no Anthonio Cicestrien' episcopo, venerabilibus viris Edwardo Stanhope, Daniele Dun, et Richardo Swale, legum doctoribus, Will'mo Barlow, presbytero, sacræ theologiæ professore, capellano dicti reverendissimi patris, Johanne Parker et Edwardo Aylworth, armigeris, Michaele Murgatrode, Georgio Whitegifte, et Georgio Paule, generosis, Will'mo Thornhill, presbitero, in artibus magistro, capellano dicti reverendissimi patris, Johanne Scott, Abrahamo Hartwell, Christofero Wormeall, et Richardo Massinger, generosis, et multis aliis in numero copioso congregatis. Super quibus, &c.

THOM. REDMAN,
Notarius Publicus, antedictus.

No. XIII.

A Forme of givinge my Almes-men their Roomes.
(From a MS. in the Lambeth Library, No. 275).

JOHANNES, Providentia divina Cantuariensis Archiepiscopus, totius Angliæ Primas et Metropolitanus: Dilecto nobis in Christo A. B. salutem in Domino sempiternam. Debilitatem tui corporis, paupertatem, et senium attendentes, locum & allocationem unius pauperum Hospitalis Sanctæ Trinitatis in Croydon, ex fundatione nostri Johannis Whitegifte, Archiepiscopi Cantuariensis, tibi ad terminum vitæ suæ, & ad sustentationem tuæ paupertatis, concedimus per præsentes; statuta & ordinationes ejusdem hospitalis volentes & te firmiter injungentes custodire, & in omnibus observare. In cujus rei testimonium, &c.

No. XIV.

Eight Letters from the Rev. Samuel Finch, Vicar of Croydon, relative to Whitgift's Hospital. (From the Originals, in the Lambeth MS. Library, No. 275).

I.—*To the moste reverende Father in God my verie good Lorde the Archbushop of Canterburie, his Grace at Lambith, with speed.*

My humble duetie remembred unto your Grace. Yesterday, being Thursday, Wolmer the bricklayer was here to vewe your worke. And he sayeth that he cannot be here himselfe: but he wyll appointe one from Westminster to be here, who will not come under xviii đ. the day, and his laborer xii đ. Hyllarie sayth, he canne bringe one presently whome he knoweth, and will warrant to take the charge, and discharge it with credit, for xvi đ. a day; and laborers we canne have inowe: thers v li. a yeare saved in ii đ. a day wages. And beside the master workeman muste be here still to conferre with the carpenter. Thus muche Hillarie tolde me; but he knoweth not of this intelligence unto your Grace. The yarde ys all defenced in, strong and saffe. This day we make an ende in pullinge downe as yet. Nowe we take morter-makinge in hande, clensinge and leavellinge of the grounde; and by Monday come sevenighte, Hillarie saith, we shall be readie for the foundacõn and bricklayer. Weeks the bricklar hath bene at your brick-clamps, and commendes them for verie good. We have our sande from Dubbers Hill: for the Parke fayleth. Thus muche I thought good to signifie to your Grace; and I pray God prosper the worke, and blesse your Grace with health to see it in prosperitie to Gods glorie. Amen. From Croydon, this Fryday the viiith of Februarii, A° 1596, R. R. E. 39. Your Graces in all duetie bounde,

SAMUELL FINCHE, *Vycar.*

II.—*To his assuered and verie lovinge Friend, Mr. Woormall, at Lambeth.*

With my verie heartie commendacons to yourselfe, Mrs. Wormall, and my wyfe, and the like from her daughter to you all, with as hartie thanks to Almightie God for Mr. Comptroller's dissolucon from the bondage of his corrupte bodie, into the glorious lybertie, noe doubte, of God's children. Sir, assure yourselfe, I forget not that it is meete that his Grace beginne the foundacon. But yt will not be readie for his Grace tyll Monday come sevenighte. By Hyllaries choyce, one Henry Blease and John Greene, bricklayers, and my parishioners, have joyntlye taken the charge of the bricklayinge worke, and have xvd. apeece the day. Blease hath begunne the groundworke nexte the highway leadinge to London; and findinge that grounde made and false, digged the trenche alonge the door unto some iiii foote deepe and iii foote wide, and ware little or nothinge combred with water; and findinge firme grounde, they have filled up that trenche with great flinte and small stone, and brickbatts, and rubbushe, not confusedly, but orderlye layed in, and rammed stronglye, course upon course, stronge and sure. This trenche revomed those small stones that lay in the court yarde, which his Grace made the boyes gather out of the church yarde, and some halfe dosen loads of small stone fetched out of Smithdoune bottome, which were there redie gathered the last yeare for the highwaies; and from thence we fetche still and lay by; the same receaved also the moste parte of those stones his Grace did see in the yarde there. We have also provided cartes to fetche us great flinte and chalke for the buildinge, and small for fillinge: because the lower grounde is not soe good and firme as the upper, and the waite of the worke may not be trusted only uppon brycke; and four loads of flinte, which come to x s̃. will well save one thousand of bricke at xvi s̃. I need not tell you that I shall lacke monie for this weeke, because the bearer herof ys Wm. Tagburne, who had v li. of me this morninge to bye two horse tomorrowe in Smithfeilde. I knowe he will tell you of it, and therefore you need say nothinge therof. The laborers have digged up iiii skulls and the bones of deade persones in the trenche that they are nowe in digginge, nexte the highway leadinge to the Parke. Thus we woulde be glad all might be well to his Grace's good lykinge. And soe fare you well. From Croydon, this Thursday morninge, the xviiith of February, 1596.

Yours as his owne,

Samuell Finche.

III.—*To his assuered and verie lovinge Freinde, Mr. Wormeall, at Lambith.*

With my verie heartie commendac͠ons, I received this morninge of Wm. Tagburne xxli. from his Grace, as appeareth in my note. And, God willinge, his Grace shall not be defrauded in stuffe, worke, or wages, as long as I have the lookinge thereto. For the skulls, there were iiii digged up indeede; and I presentlye upon the findinge of the firste did conferre with Outred, and asked him yf his conscience were cleare; and he sayd, that yt was cleare. I reasoned also with Morris, an old Welchman that had dwelt there a longe tyme, and he knewe nothinge. Moreover, for a better satisfac͠on in this matter, I caused Hillarie to caste the measure of the grounde this day. And we finde that the bodies coulde not lye within the compasse of the howse; for (to the ende that the plotte might be caste square) there was v foote taken in of the way againste the George, and iiii foote lefte out of the grounde (whereon the howse stood) againste the Crowne (as Mr. Doctor Bancrofte knoweth well). Soe that the skulls being in the trenche nexte to the George, Hillarie dare depose they ware without the compasse of the howse. Besides, there be manie that canne remember, when they digged in the middest of that streate to sette a maypoale there, they found the skull and bones of a deade person. Soe that it is generallie supposed that yt hath bene some waste place wherin (in the tyme of some mortalitie) they did burie in. And more I cannot learne.

I thanke God, our groundeworke is greatlye commended of all that vewe the same. And I hope well that will like his Grace at his comminge; for yt is not slubbered uppe, but strongley donne. I pray you give your wyfe hartie thanks for my wives curteous entertainemente, lodginge, &c. And I doe thanke you bothe for the same. My wife commends her to you. She is not verie currante yet. Thus I cease. From Croydon, this xixth of February, 1596. Yours as his owne,

SA. FINCHE.

I sende you here the copie of the condicõn of the Free-masons bonde.

Nicolas Richardson and Christopher Richardson, citizens and free-masons of London, and Gabriell Anscombe of Charlton, yeoman, are bounde to Samuell Finche, John Kinge, and William Tagburne, in c pounds. The bonde beareth date the xix of February, A.D. 1596. R. R. 39. And here followeth the condicõn.

The condicõn of this obligacõn is such, that if the above bounde Nicholas Richardson, Christopher Richardson, and Gabriell Anscombe, they or anie of them, do bringe, or cause to be broughte, to that place of the foresaid Croydon where his Graces hospitall is in buildinge, soe muche good and seasoned free-stone as shall be sufficiente for those dores and windoes belonginge to the said intended hospitall as shall be made of free stone; and shall worke the same, and sette them up, in suche necessarie and redie manner as that the worke or buildinge be not stayed or hindered through there defaulte; the dores being wrought fayre and comelie as suche dores ought to be, and the windoes with bowge worke; bothe dores and windoes of a lawfull, substantiall, and sufficient syze, in suche forme and sorte as no workman shall justlye reprehende or finde faulte with either stuffe, workmanship, or size; receivinge or takynge for the said stuffe, provision, bringinge, workinge, settinge up, and full finishinge of the same, onelie ix d. the foot for the windoes, and x d. the foote for the dore cases: Than this present obligacōn to be void and of none effecte, or else to stand and abide in full force and vertue.

Hereunto (as the manner is) they have sett there hands and seales, the day above written, and delivered the same in the presence of Antonie Bickerstaffe, George Miles, and others. Every one to have v li. in hande, viz. on Satmonday next, and v li. more when they have brought in x li. worthe of stuffe; and after that to be paid as they shall furnishe and finishe. Moreover, for the preservacōn of the groundworke, we have agreed with them to make the water table on the foresides for vii d. ob. the foote, and the crests as hiegh, for the safegarde of the windoes, for viii d. the foote. Dated the xxii of February.

Yours, SA. FINCHE.

IV.—*To his assuered and verie lovinge Freind, Mr. Woormall.*

With my verie hartie commendacōns from myselfe and my wyfe to yourselfe and your wyfe: with the like thanks for all the courteous enterteinemente you bothe shewed unto my wyfe. Syr, soe it ys, that this good-lye seasonable weather, as it hath staied our worke somewhat this day, soe it do the cause that we shall not be readie for his Grace this nexte weeke, viz. untill Monday come sevenighte, for this weather wyll not serve for layinge of morter. Neverthelesse, we doe goe on with the groundworke.

First, we have finished the two trenches next the Crowne and the George, and made them even with the ground.

Also the ynner trenche, which doth countermaunde those other, we have filled and finished on that side next the Crowne. We have digged the other that answers that against the George, and we have almost filled it this afternoone (for feare, if the weather breake, it mighte fall in againe). And whereas bothe these ynner trenches doe meete with there angle in the sellar, we have made up that angle from the bottom of the sellar, wall-wyse, with stone and morter, almost even with the grounde, and are now fillinge the voyde rometh therin with earth and rubbishe. This beinge done, we meane to goe in hande with other groundworke, untill the wether serve to worke above grounde; and order our businesse soe to the tyme, that this kinde of weather shall not hurte us, and lyttle hinder us. Thus I commit you to God. Croydon, this xxvith of Februarii, A° 1596.

Yours as his owne,

SAMUELL FINCHE.

V.—*To his verie lovinge Freinde, Mr. Wormall, at Lambith.*

Sir—With my verie hartie commendacõns to yourselfe and to Mrs. Wormall, I did understande by Mr. Mylles, that (upon on Blease his complainte) he had moved my L. Gr., as though it ware needfull that our workmen-bricklayers shoulde be loked unto (not as a caveat for us, but as a reproche to us that be overseers) as though there ware some unskilfull admitted alreadie. This Blease is one of those whome Hillarie chose with Greene to be those that shoulde take the charge of the bricklainge; and in that respecte he is allowed, as Greene ys, a penie in a day more than an ordinarie workman. Nowe, yf this Blease had had a farther insight into mens works than his partener, it had bene his parte to have made it knowen to us that are overseers, and not to have moved the matter to Mr. Mills. But shall I tell you? When these two ware chose by Hillarie, Blease begins to take a pride in himselfe, as one that woulde challenge or thought himselfe worthie of the cheifetie of all, and begins to complaine to me againste Hillarie, because he taks upon him both to sette out the bricklayers worke and give his advise for the workmen. "For," saith Blease, "I knowe better what belongs to our worke than he; and yf I be appointed one to take charge, 'tis reason I appointe the worke and workmen." I, perceivinge this, persuaded Blease to be contente to suffer Hillarie to have an insight into all mens doings. "For," said I, "the charge principallie ys his for all; and as he hath put you in, soe, yf you contente not yourselfe, he may put you out: because, whosoever commeth in here as bricklayer or bricklayers must be one with him. But, goodman Blease," said I, "I doe understande that you shoote at another matter, which neither you, nor Hillarie himselfe, nor never a man here shall atteine, if I can knowe yt; and that is, you woulde have the appointement of the workmen under you, to make a gaine of ther wages: as, for exemple, here is Kilnar, a bricklaier, one commended to us by Rowland Kilnar, his Gr. servant, a good workeman; he hath xiiii d. a day of us, and you have made him promise you ii d. a day out of it, pretendinge that he is under you, and commeth in by you, when you give him neither meate, drinke, nor lodginge; and thus you woulde doe with others: but you shall not have your will; and if he be meete to serve you for xii d. a day, he shall serve my L. soe: yf not, tis noe reason you shoulde gaine by his worke to my L. losse; for I have learned the tricke of you all; when you gaine by them, you suffer them to worke at pleasure; but if you knowe the contrarie, than you haste them on." After this, Blease seemed to be verie quiet (as it seemeth, not contented); for Hillarie and myselfe told him, if he woulde not be quiet, that i d., which he hath in the day more than another, shoulde taken from him. Indeed, hither came from Lewsham one Johnson, upon Monday was a sevenight, and did thinke to have bene imployed as a workman; but we, learninge what his skyll was, did not suffer him; yet Blease, by his leave, as carefull as he pretends to be, did suffer him for an hower, till Hillarie spied it, and woulde have suffered him as under him, but we woulde not; and then the fellowe wrought iii days as a laborer, and had ii s. vi d., a laborer's wages, as apeareth in the week's accompt which you had last. Two other came also on Friday last from Lewsham, and pretended they ware workmen, and set on to the wall that was made out of the sellar; but one of them proved none, and was paid as a laborer, for a day and an halfe, xv d., as apeareth in the accompt, and soe departed. Why? what are these matters to troble my L. with? We shall have ynough hereof yer the worke be ended, as I told Mr. Mills. Tis noe caveat to mee: for I knowe in a multitude there will fall out suche matters. "We," said I, meaninge myselfe, my father Hillarie, and William Tagburne, "will and doe joyne together as one, for the furtheraunce of his Gr. worke; and if we cannot appease, we will thruste out unrulie persons." And I pray you hartelie, Mr. Wormall, acquaint his Gr. with these my letters, as in your discretion you shall finde best opportunitie. Soe fare you well. Nowe this harde weather we get in carriages of stone and bricke, and make redie our chalke-pitts, and meddle not with other worke. We cannot as yet bargaine with a brickmaker, neither will we unadvisedlie. We will see the worke goe on, and howe our owne may serve. *Iterum vale.*

Yours, SAMUELL FINCHE.

Croydon, Marche 3.

VI.—*To my verie lovinge Freinde, Mr. Wormall, at Lambith.*

With my verie hartie commendacons, &c. Rednap came hither this day; and assone as ever he came into the yarde, and sawe the bricks, his harte was deade. He went to them, and chose here one, and there, and knockt on it, and said, "he hoped there war better to be founde in the Parke." To the Parke we came, and there wente from clampe to clampe; and here he found and there some one or moe good, but not to the purpose of his owne expectation. Fain he woulde have excused himselfe, but his handieworke spake against him; and we ware soe rounde with him, that he burste out into teares, sayinge, "he was never the lyke served in anie worke (he was ashamde of it); he coulde not excuse it; yt was the wickedness and deceitfulnesse of the yearth. And albeit he coulde not thoroughlie make amends, yet he coulde be contente to doe what lay in him; but not of that yearth." Well, than, to the lome-pitts beyond Dubbers-hill we came, neere Halinge-gate (where bricks had bene made in tyme past). There he founde suche moulde as contented him, and with much parlinge was contente to give my L. the makinge of fiftie thousande, and of x thousande for waste, (nothinge in comparison, but yet as much as we coulde get him to yealde unto), and to make l thousande more at the price he made for in the Parke, having all necessaries founde him as he had in the Parke. And there, wood must be had of from the farme-grounde, and water fetched in a carte from the other Halinge-gate. And these bricks shall be readie for us before Whitsontide. Only he requested his Gr. letters to Sir John Box (in whose worke he is) that he will be contente to spare him till he served our turne, which he knoweth he both may and will. And howe all this may be accepted of, he lokes for present answere.

Besides this, you shall receave of this bearer a paterne of the hospitall-gate, from the Free-masons; and by this paterne, vewe may be taken where his Gr. armes shall be placed, and where the dedication S. TRINITATI. There is space one eache side for VINCIT QUI.——PATITUR, &c. And for a enteringe stone of eache side, one with the armes of Woster, and the other what else is thought good.

I pray you, let the armes be drawen out in suche full proportion as his Gr. will have them, and the place sett down where, and the inscripc̃ōns what, that all things may be to his Gr. best likinge. And this must be returned with the paterne by Saturday nexte*. And soe I commit you to God. Croydon, this vii of Marche, 1596.

Yours as his owne,

SAMUELL FINCHE.

I understand by your letters you remember to sende monie.

* The following directions were sent, by way of answer, from Mr. Wormeall:—

For the Foregate of the Hospital at Croydon.

The armes over the doore must be without helmet and mantelling, and must be the armes of the See of Canterburie, viz. the Pall in pale, with the nowe Archebyshopp's armes, and the yeare of the Lord under them, viz. 1597. Over the said armes a free-stone square, with theis words in great letters, viz. SANCTÆ TRINITATI SACR. On the bare places over the gate, called (as I thinke) the Ashler, this sentence following to be written in great capital letters, viz.—

PROV. 28. QUI DAT PAUPERI, NON INDIGEBIT.

VII.—*To his assuered and verie lovinge Friend, Mr. Wormeall, at Lambith.*

Mr. Wormeall, with my hartie commendacōns to yourselfe, with praises to God for Mrs. Wormeall's good recoverie, &c. I received this morninge your letter, which doth satisfie me well, both for Birk's matter and inquirie about his lease, &c. Sir, I returne you heare his bill of suche charge as his Gr., of his owne gracious motion, wylled shoulde be imployed upon the chauncell called the Bushop's Chauncell; which war soe donne. And I sawe the leade weighed to and fro. In his Gr. note I doe not give his chauncell that name, onelie because this his doinge shall be noe president of claime hereafter; and soe I pray you shewe his Gr. For as the parishes reparacōns are registred in there churche booke, soe will this. And therfore I will loke to it, that it breed no prejudice hereafter. And thus fare you well. Croydon, this xxviiith of October, 1600.

His Gr. charge of his chauncell cometh to xxxvii s. ix d.

Your lovinge freind,
SAMUELL FYNCHE.

VIII.—*To his assuered and verie lovinge Freind, Mr. Wormeall, at Lambith.*

Mr. Wormeall, with my verie hartye commendacōns, I send you here inclosed an accompte of the voluntarie charge his Gr. hath bene at this yeare, in repayringe the chappell of Croydon churche, which is nexte to his mannor there. I sende you also a note of the whole charge his Gr. hath bene at about the same, both last yeare and this. For the accompte, I thought good not to make it with the accompt of the hospitall, because the worke ys dyvers; albeit I have more monie of his Gr. in my hands than this cometh to. For the note, I sende yt because his Gr. may knowe what the whole charge ys that he hath bene at that way. And I may tell you, for that Mr. Weller tolde me, that Robert Jones movinge my L. Admirall about his contribuc͠on to the repaire of our churche, he should aske What my L. of Canterburie gave; soe I acquainted Mr. Wellar with my L. his charge, to see if his Gr. example will drawe on anie other. Moreover, I pray you shewe his Gr. that mother Dyble, one of his Gr. pore in his hospitall, ys dead this laste nighte; her allowance of iis̃. a weeke ceaseth. Albeit, Margaret her daughter is in good hope to supplie her mother's romthe, at least for her abydinge there, whiche (as I tell her) I cannot promise her, untill I knowe his Gr. pleasure. Thus I cease to troble you anie farther. From Croydon, this xviiith day of November, 1600.

Yours as his owne,

SAMUELL FYNCHE.

No. XV.

Visitatio Hospitalis S'te Trinitatis in Croydon, authoritate reverendissimi X'to patris Gulielmi archiep' Cant', in capella sive oratorio ejusd', 11 Aug' 1634, coram d'no Edmondo Scott, milite, et Samuele Bernard, S. T. B., commissariis d'ni reverendissimi patris, inter horas nonam et undecimam ante meridiem ejusdem diei, in præsentia mei Sacvili Wade, N. P. (Ex Reg. Laud, fol. 206 a).

Articles ministred by the most reverend Father in God William Lord Archbishop of Canterbury, his Grace Primate of all England and Metropolitan, to the Hospital of the Holy Trinity, in Croydon, Aug. 11, 1634.

1. Imprimis, Whether the said lord archbishop is, and hath been, by your founder, and by letters patents under the great seal of England, appointed and authorized visitor of your hospital, and hath power to punish such offences as are contrary to the statutes and ordinances of the said hospital and the founder's intention? and hath also power to injoyne unto you orders for the good of your hospital, as often as his Grace shall see cause?

2. Item, What are the yearly revenues of the said

hospital, with the woodsales, and all other extraordinary receipts?

3. Item, What are the ordinary charges that go out thereof *singulis annis*, and what extraordinary?

4. Item, How many loads of wood are yearly, one year with another, felled on grounds and lands belonging to the said hospital; and to what uses was and is the same yearly converted?

5. Item, How many beds are there for the poor of the said hospital, and what other goods, household stuff, and utensils of household, are there in the said hospital, and thereto belonging?

6. Item, Whether there be belonging to the said hospital a common chest, to keep all the donations, charters, and evidences of the said hospital in?

7. Item, Whether there be a perfect terrier of all such lands and possessions, and an inventory of all such goods as belong to the same hospital?

8. Item, What leases there be made of the same possessions, and to whom they be made, and by whom they were made, and when? and for how many years, or what other terms?

9. Item, What fines have been taken for the said leases respectively, and by whom; and whether the same have been wholly employed to the use of the said hospital; or whether any part thereof have been employed to the private use of some other, and of whom?

10. Item, Whether any goods, moveable or immoveable, appertaining to the said hospital, are sold away? and when, and by whom, and for how much, were the same sold, and to whom?

11. Item, Whether have the poor of the said hospi-

tal their due allowance, according to the ordinances and statutes of the said hospital, and as they ought to have—as meat, drink, lodging, and apparel; and if not, by whose default is it?

12. Item, Have you, or any of you, taken any money for admittance of any the poor men, women, or children, into the same hospital, or for procuring them so to be admitted?

13. Item, Whether the master, warden, schoolmaster, usher, or any of the almsmen or officers of the said hospital, have offended against the statutes and ordinances of the said hospital; and when, and wherein?

14. Item, Whether the schoolmaster and the usher perform their duties, in instructing the youth committed to their charge? and whether is the schoolhouse and schoolmaster's house kept in such repair as is fitting? and whether do the schoolmaster and the usher carry themselves sober and free from scandal, as the statutes require? and whether doth the schoolmaster duly read divine prayers in the hospital chapel, as is required?

15. Item, Do any of you know any thing concerning the said hospital, or any part or member thereof, that is fit to be amended? declare it, and free your consciences.

No. XVI.

Heads of Orders for the Charity Schoole for ten poor Boys and ten poor Girls, founded by Thomas Tenison, Lord Archbishop of Canterbury, at Croydon, March 25th, 1714. (From a MS. in the Lambeth Library, No. 806. 5, intituled "Croydon School Orders").

1. This schoole is to consist, at present, of a schoolmaster and mistress, Mr. Henry Zealy and Mary his wife, who shall teach no other children but what belongs to this school, namely, ten poor boys and tên poor girls.

2. The master and mistris shall always be profest members of the church of England, of sober life and conversation; either to be twenty-five years of age at the least; they must frequent the holy communion, and understand well the principles of the Christian religion.

3. The master shall be able to write a good round hand, and understand the grounds of arithmetick, and teach the children the true spelling of words, with the points and stops to true reading.

4. The master shall, twice a week at the least, instruct all the children in the church catechisme, and by some exposition approved of by

5. No boy or girl to be under eight years old when admitted, nor to stay till after fourteen, unless it be to even the quarter then going on.

6. Each boy and girl to be sent in cloathed whole and clean.

7. A Common Prayer Book and Bible to be provided for each boy and girl.

8. The boys are to be taught to read, write, and arithmetick; the girls the same, also to spin, knit, sew, and work.

9. They are to come to school in the summer at seven in the morning, and stay till eleven; to come again in the afternoon at one, and stay till five. Summer to be reckoned from Lady-day to Michaelmas.

10. In winter to come at eight in the morning, and stay till eleven; to come again in the afternoon at one, and stay till four.

11. Every Lord's day and every holiday, and every Wednesday and Friday, they are to go to church two by two, to set orderly in their proper seat, and make the answers at the prayers, and sing the psalms.

12. They are to break up at Christmas, Easter, and Whitsuntide, and have the usual liberties as at other schools.

13. On Thursday they are to leave school at 3 of the clock in the afternoon.

14. On Saturday in the afternoon, the girls, five at a time, in such order as may be most easy, are to help to clean the house.

15. Absence from school, or great crimes, as lying, swearing, stealing, prophanation of the Lord's day, shall be noted in weekly bills, to be laid before the trustees at their meeting, in order to their correction or expulsion.

16. The mistress shall weekly chuse one girl to be her particular assistant for the week.

17. If the parent, brother, or sister of any one in the school shall steal any thing from Norwood, the child related to them shall immediately be expelled, and forfeit the school cloaths and books.

18. If any one of the children of the school shall leave or be taken from the school before they have learnt what the statutes shall direct, the cloaths and books belonging to such shall be left for another.

19. If the parents or friends send not the children clean, decent, washed, and combed, or not at the school hours, or any ways hinder them from observing the orders of the school, such children to be dismissed.

20. No child whose parents frequent the meeting-houses shall be admitted, or continue if admitted.

21. The trustees are to meet on the Tuesday after every quarter day, to look into the state and condition of the school; and then these orders are to be read publickly before them, the master and mistress, and all the children.

22. The trustees shall pay unto the master and mistress each pounds quarterly, out of estate purchased by the founder, Thomas, Lord Archbishop of Canterbury, for the perpetual support of this very charity schoole.

23. It shall be lawfull for me, Thomas, the now Archbishop of Canterbury, founder of the said school, to abrogate, add unto, change, or alter these orders, to place or displace any part thereof, and wholly to go-

vern the same, according as shall to me seem reasonable, during my natural life, without any other person intermeddling therein.

24. The school master and mistress to sit rent-free in the new school house purchased and fitted up by his Grace, Thomas, Lord Archbishop of Canterbury, situate in the parish of Croydon.

25. No child to be admitted whose parents are not legally settled as inhabitants of the parish of Croydon.

26. At their coming in the morning, the master, or one of the scholars appointed by him, is to begin with the prayer, " Prevent us, O Lord, in all our doings," &c.; then the collect for the day; and then the collect for the fifth Sunday after Trinity, " Grant, O Lord, we beseech thee, that the course of this world may be so peaceably ordered," &c.

27. At night, at their going away, they shall say the collect for the day, and then the collect for the fourth Sunday after Trinity, " O God, the protector of all that trust in thee," &c.; and also, " Lighten our darkness," &c.

They must be charged when they go to bed to say (as in Psalm 4th, unto verse 9th), " I will lay me down in peace and take my rest, for it is thou, Lord, that makest me dwell in safety."

No. XVII.

Acts of Parliament.

For repairing Croydon Church—1 Geo. 3, c. 38.

For selling the Palace—20 Geo. 3, c. 57.

For inclosing the parish—37 Geo. 3, c. 144.

For a canal from Croydon to the Grand Surry Canal—41 Geo. 3, c. 15, and 51 Geo. 3, c. 11.

For inclosing the waste lands—43 Geo. 3, c. 53.

For re-building the Court House, &c.—46 Geo. 3, c. 130.

For making a road from Foxley-hatch, in Croydon, to Riegate—47 Geo. 3, c. 25.

No. XVIII.

The Case of the Inhabitants of the Town and Parish of Croydon, in the County of Surrey, concerning the great Oppressions they ly under, by reason of the unparalleled Extortions, and violent, illegal, and unwarrantable Prosecutions of Doctor William Cleiver, Vicar of the said Parish: Humbly presented to the Consideration of Parliament.

[First printed in 1673].

The said Doctor William Cleiver, in the times of the late Rebellion, obtained a sequestered living, called Ashton, in Northamptonshire, in which he behaved himself much unlike a clergie-man, as will appear by the articles annexed, the which were in those days exhibited against him. However, there he continued to persecute the poor people till some time after his Majesty's most happy restoration to his crown and dignity; when Doctor Whitford, the person sequestred out of the said living, being about seventy years of age, and living, was restored to his benefice. When Cleiver got this living, he entertained one Mr. Preston to be his reader, who accepted thereof, served and officiated there in that capacity; but Cleiver would never pay him his wages; so that he might have starved, if some of the parishioners (to whom by stealth he did sometimes read common prayers and divine service,) had not given him relief; for which Cleiver caused him to be sent for up to London, by a messenger; and, being so old that he could not ride on horse-back, he was brought up in a cart stufft with straw, and kept at London till utterly ruined; and then they released him.

The articles exhibited against the said Cleiver by the said parishioners were as follows:—

Articles exhibited against William Cleiver, Minister of Ashton, in the County of Northampton.

First, That the said Cleiver is a very covetous man, and doth endeavour unjustly to exact and extort sums of money and other things from the said parishioners, and others that he hath to deal for.

The said Cleiver did unjustly demand a cow for a herriot, of Sarah Honor, a poor widow, whose husband was lately dead, and died so poor that the parish was forced to bury him at their charge, he leaving his said wife with five small children, and nothing to maintain them but that one cow; and it was never known that any herriot was there paid. And, the better to procure his end therein, the said Cleiver promised her, that if she would let him have the cow, he would procure her a warrant from the justice of the peace, that the parish should pay her weekly one shilling, for the maintenance of herself and children.

The said Cleiver hath several times unjustly detained and withheld the wages of labourers from them, who had painfully and faithfully done their duties.

He caused a poor widow (whose husband was then lately dead, and she herself being then very sick and weak, and almost blind, and left in debt fifty pounds at least, having four small children, unable to maintain themselves,) to pay him the tenth penny for a calf, which she sold towards the payment of her husband's debts, and enforced her to pay two-pence for the head and pluck, which she had reserved for herself.

The said Cleiver hath pulled down a great part of the parsonage-house, and converted the materials thereof to his own use.

The said Cleiver, notwithstanding his parsonge at Ashton is worth one hundred and twenty pounds by the year, at least; yet, for the lucre of money, did undertake to serve another cure of twenty pounds by the year; by means whereof he served neither of them as he ought to do.

The said Cleiver is a very contentious man, and doth

much vex, and trouble, and disquiet his neighbours, parishioners, and others, by unjust suits and malicious troubles.

The said Cleiver did, in a clandestine way, procure a bond which was made to a sergeant at arms for security, and sued one Budworth, who was surety herein, upon the same, and recovered fifty pounds of him, and caused him to spend fifty pounds more, at least, notwithstanding the sergeant of arms was never damnified one penny thereby.

The said Cleiver hath, at the last assizes in the county of Northampton, indicted a gentlewoman of his parish, of good parentage, and of worth and quality, for felony, for stealing of his horse; and himself and his wife and maid gave evidence thereupon, and would have brought the gentlewoman's life in question, if he possibly could have done it; and, when the grand jury had returned the bill Ignoramus, he gave out in speeches, that he had sufficiently disparaged the gentlewoman by what he had done.

That he hath commenced suits against divers of his parishioners who were willing to pay him his just dues, because they would not pay what he would unjustly have exacted and extorted from them.

He hath refused to pay just debts owing by him, and hath declared, he would rather spend one hundred pounds in law, than pay five pounds when he is sued for it, though it were due.

The said Cleiver is very weak, unable, and insufficient for the ministrial function, and idle, and will not take pains therein as he ought to do, nor perform what of duty he ought: For,

He doth ordinarily preach other men's works, *verbatim*, that are in print, which the parishioners have in their houses, and can read at home.

He hath several times in his sermons uttered and spoke nonsense.

He hath neglected his cure upon several Lord's days, to prosecute contentious suits and quarrels, and for his own ease, and procured none to supply the same.

Refused to baptize the child of a visible believer, being tendered, and the father present.

Refused to baptize the child of a soldier that was in service, because the father was not present; and the child died within few days.

The said Cleiver is scandalouse in his life and conversation. For,

First, He is a liar, and a common speaker of untruths.

Secondly, He did privately keep in his house one Mistris Bernard, widow, six or seven weeks together, and denied that she was there; and afterwards kept her publicly, having no relation to her, and now keepeth her as his wife; but whether they were ever married according to the laws is not known.

That he the said Cleiver hath been drunk, and abused himself with excess.

The King's Majesty being restored, the sequestred incumbent, Doctor Whitford, entered upon the said living; and Doctor Cleiver, being thereupon to seek for a benefice, came to London, and then pretended himself to be a zealous son of the church (though for ten years before he had possest the sequestred living, and violently prosecuted his reader, for reading common prayer, as aforesaid); and, not being well known about London, hearing that the vicarage of Croydon was void, made friends to the Earl of Clarendon, then Lord Chancellor of England, to obtain the same; which he effected by the help of a gentleman (to whom he promised a good reward for his pains, to be paid so soon as the presentation should be sealed); which gentleman, having obtained the presentation, the Doctor got it from him, but never paid him to this day, as the gentleman reports.

That, having got this presentation of Croydon, he hastened thither. The vicarage consists only of small tythes, which at the utmost value is not worth above eighty pounds. This, for some time, was paid the Doctor *per annum;* who, when he had been a little in the parish, and had got all the parishioners' names into his book, fell to his old practices of oppression and extortion, bringing frequently vexatious suits against all or most of the parishioners, because they would not comply with his unconscionable and extravagant demands. Under these horrid oppressions the parishioners having many years suffered, and some hundreds of the inhabitants ruined thereby, they joined together, and caused the said Doctor to be indicted for a common barrater; which was to have come to a trial in Hillary Term, 1673, at the King's Bench bar; accordingly, the jury was summoned, and the informers ready to have made good the indictment by above three hundred witnesses; but, about two daies before the said trial should have been, a *Noli Prosequi* was entered, by means whereof the trial was stopt.

This *Noli Prosequi* was obtained by fraud also; for Cleiver, finding that the parish were resolved for to prosecute, came to one Mr. Bickerton, who was then one of Mr. Baron Turner's clerks, tells him thereof, adding, that he was like to be undone thereby, for that the rogues (as he called them) would swear him to be a common barrater; then asked him, whether he believed there could be such rogues in *England?* To which the said Bickerton replied, that he always suspected that a mischief would come upon him for his continued vexations continued to his poor parishioners, and wondered they had not done it sooner; adding, that he believed they had just cause to proceed against him, by what he understood of his ways in proceeding against them in the Exchequer. Whereupon, Cleiver asked if there were no way to stop the trial for that time; begged his assistance, if possible, to do the same, promising to give him ten pounds for his pains, if he succeeded therein. Upon which, Mr. Bickerton advised him to get a *Noli Prosequi;* for which purpose, a petition was drawn, setting forth, that the parishioners were litigious, factious people; would pay him no tythes; but forced him, by means thereof, to sue for his dues; and now had indicted him for a common barrater, merely for bringing such his lawful suits against them; therefore prayed proceedings might be stopt. This done, the said Doctor gave Bickerton a note under his hand, to pay him ten pounds if the suit were stopt that term; and away he went with the petition, to a person that presented it to his Majesty, who (being misinformed, and thinking that the said Doctor might have just occasion to sue, as by petition he pretended), did order Mr. Attorney-General to enter a *Noli Prosequi;* and accordingly, a *Noli Prosequi*, reciting his Majesty's command, was granted, and the suit stopped; which done, the said Doctor went back to Bickerton, and, crying, told him his Majesty denied the petition, and that the trial was not stopt; desired, therefore, his note again, that he had given him for his ten pounds; which Mr. Bickerton delivered, as thinking Cleiver had spoken truth; and so cheated him of his ten pounds promised him upon accompt as aforesaid.

The poor parishioners of Croydon, they were greatly troubled, not knowing what to do, being stopt from proceedings at law, thereupon went to counsel, and were advised to have petitioned the Parliament, then sitting, as lying under a grievance, and being denied the benefit of the law for their redress; but, upon application made to the Lord Keeper and Mr. Attorney-General that now are, and acquainting them with the truth of the case, they did assure the parishioners, that his Majesty was surprised with the granting of his *Noli Prosequi*, and undoubtedly would take the same off, if addressed unto the Council.

Whereupon, the 21st of March, 1672, a petition was exhibited to his Majesty, praying that the *Noli Prosequi* might be taken off, and the parishioners left to the law; or, that he would be graciously pleased to hear the cause, and relieve them according to justice.

To which petition was annexed the articles following:—

Articles of high Misdemeanour, humbly exhibited to the King's Most Excellent Majesty, and the Right Honourable the Lords of his most honourable Privy Council, by the Inhabitants of the Town of Croydon, in the County of Surrey, against Dr. W. Cleiver, Vicar of that Town.

1. That the said Doctor, by unjust, vexatious, and numerous suits, by him frequently brought against his parishioners, extorts more from them than what either his predecessors claimed or had, or is his due.

2. He frequently, after he hath been punctually paid his full dues, arrests his poor parishioners, and forces them to pay the same over again; together with great sums for charges, which he pretends he hath been at; declaring he will have of them what he pleases, for he cannot live on his dues.

3. He hath served several of the poor people with pretended processes, and compels them to pay him money when there is nothing due to him from them; and extorts money for the process, when as there was never any process pursued other than what was made by himself; which is to the great abuse of his Majesty's Courts at Westminster.

4. That he doth very often sue out many writs out of the Exchequer against several of the parishioners that owe him not a farthing, puts them to vexation, trouble, and charge, and then never exhibits any bill against them.

5. That, because one of his parish would not swear for him what he would have had him, he vowed he

would sue him as long as he lived; and so hath continued to do ever since, to his almost utter ruin.

6. That he frequently arrests poor people for tithes, puts them to great charges, when as they owe him nothing; and, such as are able to make opposition he never declares against, but forces the rest to compound, and give him what he pleases.

7. That he hath attached the goods of several persons, for tithes pretended to be due to him, when as none was due from them; hereby put the people to charge, done damage to their goods; and when replevins have been brought, then he hath never appeared or declared.

8. That he hath imprisoned several persons, and detained their goods in his house till he hath forced them to give him what money they have, and seal bonds to him for other sums of money by him demanded, when there hath not been one penny due to him.

9. That, having come to an agreement with several persons for their tithes, to take such a sum for them yearly, he hath received the sum of money agreed upon, for several years together, and afterwards denied the same, and sued the persons for their tithes in kind, and the arrears thereof; and, some of the people with whom he agreed, being illiterate, trusted him to write the agreement; he set down double the sum that was agreed upon, got their mark to his book, and sued the persons, and forced them to pay the same, together with such unreasonable charge as he demanded of them.

10. He hath cited several of his parishioners into the spiritual courts for pretended crimes, because he

could not have his unjust demands of them; and when he hath so cited them, hath taken sums of money of them, to excuse them being prosecuted in the said court.

11. That, having let a lease of his tithes to one Mr. Wood, for a certain rent, Wood, by virtue of that lease, received tithes of the parishioners, and paid him his rent; after which, he sued the parishioners over again for the same tithes, and forced them, for quietness, to pay him, declaring, the only way to be quiet was to pay both Wood and him.

12. Several poor people having, in the time of the late dreadful sickness, buried relations in the woods, the said Doctor, in the time of their necessity, was so far from extending his charity towards their relief, that he forced them to pay unreasonable fees for their burials, as if they had been buried by him in the church-yard. Those that would not comply with him, he sued and extorted great sums of money from them, for his charges as well as duties, before he would clear his persecution.

13. He denied to receive his tithes in kind of several of his parishioners, though duly tendered him, and he desired to accept of them; demands of them what money he pleases in lieu thereof; and if they do not pay what he demands, he sues that at law; and if they agree and pay for one, two, three, four, five, or six years, at the rate compounded for, then, at the six years' end, he demands of them the arrears of tithes, pretending the tithes to be of greater yearly value than what he compounded for; and if the people will not pay what he demands above his composition-mo-

ney paid for three years past, then he sues them, till they, by fright, and being put to unreasonable trouble and charge, were enforced to give him what he demanded, rather than be undone by contending with him; that being the lesser evil of the two, by him put to their choice.

14. He, by this violent persecution of diverse poor men, hath forced them to leave their wives and children, and seek shelter in remote places, to the utter ruine of their families.

15. He demands of the poor inhabitants the tenth penny got by their day-labour, and threatens to compel them to bring their milk into the church porch to sell, and there deliver him his tenth. From some he hath extorted great sums of money, on pretence that he hath been at law with them seven years, whereas they were never served with any process. Others he hath served with process after they had newly paid him his demands, to put them to further vexation and charge. One process which he hath taken out against a man, he altered it, and made another man pay the charges, as if it had been originally made out against him.

16. He frequently extorts great sums of money from his parishioners, for marrying out of the parish; and those that refuse to pay him what he demands for that offence, he refuses to receive their tithes, and then sues them for tithes and that together.

17. Such persons as refuse to pay the said Doctor his unjust demands, he will not suffer the clerk to receive his just dues, threatening to sue both for his

money and clerk's dues, when as both hath been tendered, and might have been received without any suit.

18. He hath arrested several on pretended great actions, and thereby kept them in prison, and yet, on the trial, could prove nothing against them, to the utter ruin of these poor people imprisoned; and hath forced several persons to pay him five pounds, when his due is but twelve pence.

By these and the like extortions he makes his living above 250*l.* *per annum*, which never was worth, to any of his predecessors, above 60*l.*, and enricheth himself by the ruin of his parishioners, especially the poorer sort, that live on the common, whom he endeavours to enslave, because they are not able to contend with him at law.

That the petition and articles were read.

The King's most excellent Majesty, his Royal Highness the Duke of York, and eighteen more of his Majesty's most honourable Privy Council, were present; and the order following was made:—

At the Court of Whitehall, the 21st of March, 1672.—Present, the King's most excellent Majesty; his Royal Highness the Duke of York; Lord Chancellor; Lord Treasurer; Duke of Lauderdale; Duke of Ormond; Earl of Bridgewater; Earl of Northampton; Earl of Anglesey; Earl of Carlisle; Earl of Arlington; Earl of Bath; Earl of Craven; Lord Viscount Falconbridge; Lord Newport; Lord Berkley; Mr. Secretary Coventry; Mr. Chancellor of the Duchy; Master of the Ordnance; Sir Thomas Osborne.

The inhabitants of the town of Croydon, in the county of Surrey, by their petition this day at the board, humbly complaining of the many oppressions, extortions, violent and unwarrantable proceedings of Dr. William Cleiver, vicar of the said parish of Croydon, against the petitioners, under pretence of recovering his tithes, insomuch that, in a short time (if relief be not given) most of the petitioners will be forced to leave their dwellings, or be inevitably ruined, as some of them, with their families, have already been: It was thereupon ordered by his Majesty in council, That the petitioners do forthwith deliver in to the clerk of the council attending, the particulars wherewith they intend to charge the said Dr. Cleiver, to the end he may have a copy thereof timely enough to come prepared to make his defence thereunto, on Friday the 4th of April next; which time his Majesty hath appointed to hear the said complaints; and doth command that all parties concerned do then give their attendance.

John Nicholas.

According to this order, a short paper was delivered in to the clerk of the council, whereby the parishioners declared they would insist upon—

Oppression, extortion, common barratry, subornation of perjury, forgery, felony, and some more such like petty crimes of his doctorship. But, before the cause came to be tried, the act of grace came out; and barratry, a thing never pardoned before by any former act of grace, was pardoned thereby.

The 4th of April, 1673, his Majesty was graciously pleased to come early to the council; and, together with his Royal Highness, and twenty more of the Lords of the Privy Council, sat with unspeakable patience, heard the charge against the said Doctor made good; which was done by the oaths of several persons, in every particular mentioned in the said articles; and many others (if possible more violent and heinous than the former) was proved against him.

1. As, first, he, being one of the trustees of the alms-houses in Croydon, would not suffer almes-men to be admitted, till he had forced them to promise to allow him one half of the profits of the said alms-houses, and took bond for the same; which done, he admitted them; and would have turned one Edward Humfry out of his alms-house, when he could no longer afford to pay him four shillings a month, according to the said bond; which Humfry is now turned out by the said Doctor.

2. That, when as the sacrament money hath been collected for the poor, he took a third part of it to his own use, saying, *None was poorer than the vicar,* and kept it; which hath since prevented men's charity.

3. That he sent for a man from London, pretending he would pay him some money which he owed his father, who was dead. The poor man came to Croydon to his house, and was made welcome, and had good meat and drink; after which he was carried by the Doctor into the garden, made eat fruit; which done, he sent him away without a penny of money. But that was not all; for, soon after, he arrested the poor man in an action of trespass, for two and three hundred pounds damage, for coming into his ground,

kept him a prisoner till he was forced to seal him a general release, and so cheated him of his debt, which was about thirty pounds, and grew due to his father, Mr. Preston, as aforesaid, for officiating as curate for him at Ashton in Northamptonshire.

4. They proved that the Doctor had two special bayliffs, that do all his business; one of them had been burnt in the hand, the other a vile rogue as is in the country. These arrest his parishioners; then, bringing them to his house, there they are kept prisoners till he force them to what he pleaseth; which done, he draws them into judgments instead of pretended notes; and these are they whom he useth, together with a most excellent gentlewoman, Mrs. Ream's daughter-in-law, who went there by the name of Mrs. Cleiver; who, by the report of the whole country, and people at Westminster, had a child before her marriage, lay in at Westminster, and the Doctor was much with her, and she went for his wife, though her mother was then alive; now lives at the Doctor's house, wholly governs him, to say no worse; makes him oppress the people as she pleaseth, and is his constant witness to all his agreements.

5. They proved, that when the Doctor hath gone into the pulpit to preach, he hath dropt bundles of writs out of his pockets, taken out against his parishioners, and, being taken up and perused, have been found to be rased in the dates and in the names, and new dates and new names put into them; which was a frequent thing with him, he making twenty warrants out against twenty persons upon one single writ, and making each person pay seven shillings and sixpence

for charges, when the writ cost him not above three shillings; so that he got above seven pound by a writ, and made a property of his Majesty's Court of Exchequer, to the enriching himself and impoverishing his parishioners. And often, when he should have been preaching on Sundays, did use to ride to London, to follow suits against his parishioners, leaving them without any one to preach or read prayers to them.

6. That he sent for the parishioners to come to his house to pay their tithes, and when they had so done, frequently arrested some of them in actions of trespass, for coming upon his ground, thereby putting to great charge, and ruined some of them.

7. That he hired a carrier to bring his goods to London, which was done; and in London an old desk was stole out of the carrier's warehouse, by his letting his goods lie there three weeks after brought up; but the man that stole it was catched, the desk brought back to the inn, was opened, not knowing whose it was, and there was only in it a pair of slippers and one old law book. The Doctor arrests the carrier, declares against him for 190*l.*, pretended to be in the said desk; the carrier, telling his landlord thereof, he produced the said desk, with the things aforesaid in it; so the suit ceased.

8. He hath caused the gentry to leave the towne, to the ruin thereof; spoiled the school, so that no gentlemen came to it. He hath caused lands and houses to fall in their rents; brought down the price of them, in their sale, above three years' purchase; makes tenants that they will take no lease, unless landlords will covenant to secure them against him; which they dare not

do; and so the houses and lands stand empty, and lie waste. And he hath forced the parishioners to leave their parish church, and to keep from receiving the sacrament; insomuch that there are not above ten or twelve in all (beside alms-people, who are obliged), that will come to the church or sacrament; but if a stranger at any time do preach, there come at least six, seven, or eight hundred.

Lastly. That the said Doctor endeavoured to part husbands from their wives, raising false stories of them, offered himself to sue out divorces.

And many more crimes of this nature.

These matters, and also the aforesaid articles, being clearly proved, the said Doctor only for defence offered, that the petitioners would pay him no tithes, thereupon was forced to sue them for his dues; and prayed time to make defence.

Whereupon, the order following was made:—

At Whitehall, the 4th of April, 1673.—Present, the King's most excellent Majesty; his Royal Highness the Duke of York; his Highness Prince Rupert; Lord Chancellor; Lord Treasurer; Duke of Lauderdale; Duke of Ormond; Marquis of Worcester; Earl of Ogle; Earl of Ossory; Lord Great Chamberlain; Earl of Bridgewater; Earl of Northampton; Earl of Anglesey; Earl of Carlisle; Earl of Craven; Earl of Arlington; Earl of Carbury; Viscount Hallifax; Mr. Secretary Coventry; Mr. Chancellor of the Duchy; Master of the Ordnance.

His Majesty, having this day heard at large the

complaints of the inhabitants of the parish of Croydon, in Surry, against Dr. Cleiver, vicar of that place, containing several exorbitant courses by him practised for recovery of his tithes, was pleased to refer the whole matter to his Grace the Lord Archbishop of Canterbury, and to the Earl of Shaftsbury, Lord High Chancellor of England; who, requiring the said Doctor and some of the principal persons of the said parish to attend them, are to endeavour to settle the business for the future quiet of the parishioners, and that there may be a constant maintenance for the vicar there; which, if their Lordships shall not be able to effect, they are to return their opinion and advice to his Majesty, what they conceive fit to be done therein.

JOHN NICHOLAS.

With this order his Grace the Lord Archbishop of Canterbury and the Right Honourable the Lord High Chancellor of England being attended, they were pleased to appoint a day in May last past, when both parties, with their witnesses, should attend their Lordships, at Lambeth House, where the matters aforesaid were made manifest; as also—

That the said Doctor Cleiver was a notorious and common thief, that used to come into booksellers' shops and steal books, and carry them away. Several masters of shops gave testimony thereof. To which, the Doctor being to give answer, declared, that true it was that he was sitting in a shop, reading a book, and saw a gentleman come by that he had occasion to speak with, and in haste ran after him, and forgot to lay down the book, and carried it with him; but the man presently followed him, and took it of him. But, alas! good gentleman! he had often had this chance of following his friends with other men's books under his cloak. If he had done so but once, it might have passed as forgetfulness; but it fell out very unluckily, that, at Mr. Sawbridge's shop, at the Bible, on Ludgate Hill, he came in and took away a book of good value, carried it beyond Fleet Bridge, and there was overtaken and brought back, and begged pardon, declaring who he was. Mr. Sawbridge thereupon, loth to bring any scandal upon a man of his coat, went with him to Fetter Lane, to a gentleman's house, to inquire of him; and, finding that he was Dr. Cleiver, and vicar of Croydon, resolved to make no more noise of it, if he would ingenuously confess what books he had formerly stolen from him; for that he had often been at his shop, and he had often lost books, but never knew whom to charge with them. Thereupon, he promised he would go home and look over all his books, and bring him the names of them, as also money for them; that he never came there again, or sent the catalogue of the books he had stolen; but he did send his wife to the shop, with about thirty-nine shillings, and she paid the same for the books he had stolen; so that it was not always forgetfulness, or running after friends, that made him carry books out of the shops.

The said Doctor being then asked, why, after so many arrests, and so many suits against his parishioners, he never brought any one of them to trial, whereby the *modus decimandi* might have been settled; it was proved against him, that, being demanded the same question in the Exchequer, he declared, most falsely and scandalously, of his Grace of Canterbury, that his Grace had directed him not to do it, because the then Lord Chief Baron (now Lord Chief Justice Hale) was an enemy to the Church of England.

Upon the whole matter, his Grace and the Lord Chancellor proposed to the parish to settle a certain maintenance upon the vicar for the future, that so no more suits might arise. To which the parishioners most willingly assented, provided the Doctor might be removed, and have no benefit thereof; and proposed, as poor as he had made them, yet, to be rid of him, and to have a good, learned, orthodox, and peaceable man settled amongst them, they would make a certain allowance of 120*l.* *per annum*, to be paid to the succeeding vicar, by quarterly payments, without charge and trouble.

No sooner was this proposal made, but Dr. Cleiver, who, the world knows, is a most notorious liar (and the old proverb is, *a liar had need to have a good memory*), starts up; and, although just before he had told their Lordships he was forced to bring his actions, and to have multiplicity of suits, because his parishioners would pay him nothing; that, although he had offered them his vicarage at 60*l.*, 50*l.*, nay, 30*l* *per annum*, they refused to take the same; he told them that 120*l.* was nothing, for that the parish had offered him 180*l.* themselves.

Note.—The Doctor let the tithes for three years, to one Wood, at 60*l.* *per annum*, and the said Wood received the tithes of the tenants, paid him his rents, and afterwards the Doctor sued his parishioners, and made many of them, to their utter ruin, pay him the said tithes over again, though they had Wood's discharge for the same. Nevertheless, to be rid of the Doctor, the parish then proposed to allow 160*l.* *per annum* to the surviving vicar.

The conclusion of this meeting was, that the parishioners should draw up their proposals in writing, put them under their hands, and present them to the Lord High Chancellor of England. Accordingly, they were drawn up and signed by above three hundred persons, and presented to his Lordship, and a copy to his Grace of Canterbury; and one only copy, annexed to a petition, was presented to his Majesty in council. The purport was as follows:—

1. That, provided Dr. Cleiver were presently removed, the parish would allow the succeeding vicar, successively, if a sober, learned, orthodox, and peaceable man, 160*l.* *per annum*.

2. That Dr. Cleiver should have no advantage of the said offer.

3. That they would consent to an act of Parliament or decree in Chancery or Exchequer, to settle and confirm the same.

4. That, till such act or decree could be passed, to the intent the said Doctor might not be continued, twenty of the most able parishioners would become bound to the surviving vicar, to pay him 160*l.*, in lieu of all his tithes or other dues, until the said 160*l.* should be confirmed by act of Parliament or decree aforesaid.

5. That whereas the said Doctor pretended great

arrears of tithes and dues owing to him, and many actions were brought by him against the parishioners for the same, every person concerning therein would be determined by the judgment of any two indifferent persons as the said Lord's Grace of Canterbury and the late Lord Chancellor should appoint to hear and determine the same, without further charges and trouble in law.

This seemed very reasonable and satisfactory to his Grace and the Lord Chancellor; and thereupon, Doctor Cleiver, being sent for by the late Lord Chancellor, was told thereof, and advised to surrender to his Grace of Canterbury; which if he would do, was offered another living of 120*l*. *per annum*, in Northamptonshire, then in the Lord Chancellor's gift, and his Grace of Canterbury should present one other able, orthodox man to Croydon, to be named by the then Lord Chancellor, (his Grace of Canterbury being so convinced of the badness of Dr. Cleiver, as he resolved never to present him to any other); this being the medium most graciously proposed by his Majesty in council, for the ease of his poor subjects. This offer the Doctor promised to accept, and to go to the Lord Archbishop of Canterbury and resign accordingly; but, as he ever hath been false, so in this he manifested it, by breaking word both with his Grace and the Lord Chancellor; so that they could do nothing.

And truly, till it was considered what the reason might be, it was wondered that he that might have had a living of 120*l*. *per annum* certain, to part with about 60*l*. or 80*l*., gotten with contention, should not accept the same; but afterwards, it appeared plainly that he was so well known in Northamptonshire, that the people would have stoned him out of the country, if he had come thither.

Hereupon, the parishioners of Croydon once more most humbly petitioned his Majesty in council, and annexed their proposals aforesaid, shewing how inevitably they must be ruined if he were continued, and prayed his removal.

Whereupon, the 28th of May, 1673, it was ordered as followeth:—

At the Court at Whitehall, the 28th of May, 1673.—Present, the King's most excellent Majesty; his Royal Highness the Duke of York; Lord Archbishop of Canterbury; Lord Chancellor; Lord Treasurer; Lord Privy Seal; Duke of Lauderdale; Earl of Ogle; Earl of Bridgewater; Earl of Bath; Earl of Carlisle; Earl of Arlington; Earl of Craven; Earl of Carbury; Viscount of Hallifax; Lord Maynard; Lord Newport; Lord Holles; Mr. Vice Chamberlain; Mr. Secretary Coventry; Mr. Chancellor of the Exchequer; Mr. Chancellor of the Duchy; Master of the Ordnance; Sir Thomas Osborne; Mr. Speaker.

Whereas, upon hearing the complaints of the inhabitants of the parish of Croydon, in the county of Surrey, against Dr. Cleiver, vicar of the place, concerning several exorbitant courses by him practised for recovery of his tithes, it was, the 4th of April last, referred to his Grace of Canterbury and the Lord High Chancellor of England, to endeavour to settle the business, for the future quiet of the parishioners, and that there may be a constant maintenance for the vicar there. And the said inhabitants, by their humble petition this day read at the board, praying his Majesty to receive a report from the Lords' referees, of their proceedings therein, and to suspend the said Doctor, and give the petitioners leave to proceed at law against him, or otherwise to relieve them: It was thereupon ordered by his Majesty in council, that it be again referred to his Grace the Lord Archbishop of Canterbury, and the Right Honourable the Lord High Chancellor of England; who, calling all parties before them, are to endeavour to compose and settle the said business, according to the proposals made by the said inhabitants of Croydon; but, if they cannot effect the same, then it is ordered that the said matter in difference be heard at this board on Wednesday the 11th of June next; at which time all parties concerned are to give their attendance.

EDWARD WALKER.

Before the time for this hearing came, unluckily it fell out, that Dr. Cleiver, having taken a little too much of the creature, in London, being upon his journey home, just as God would please to have it (to shew what he deserved), against the gallows, near Newington Butts, his horse threw him, or he fell off from his horse, broke his leg in three pieces, and put his shoulder out. There he lay, and none would help him, the people thereabouts knowing him so well, that one cried, "There lies the vicar of Croydon, with his leg broke; I would to God he had broken his neck; the church would then be no more scandalized by such a rogue, nor the poor people tormented. Others wished more severe things, which savoured not of Christianity; therefore, we will not mention them: but, certain it is, no one would help to remove him till they were paid beforehand, because he is counted so great a knave that none would trust him; nor would a coachman take him up to carry him to Dr. Welden's house (the parson of Newington), before he had ten shillings in hand, which is not half a mile. And when he was at Dr. Welden's house, he sent for one Dr. Thorland, the bone-setter, from London, who found him in a very ill condition, but very glad he saw him at all; for it happened about twenty years since, an accident of the same nature befel him in Northamptonshire, of which the said Dr. Thorland cured him, but to this day was never paid for it. Nevertheless, the good man, as became a Christian, pitied the Doctor, and, seeing his misery, applied himself to his cure, and effected the same in about three months, when he demanded for his pains in this and the former case fifty pounds; but the Doctor would not pay him, upon which he arrested him. Thereupon, he pleaded himself his Majesty's chaplain, shewed a certificate of his being sworn in that capacity, and so got off without paying for his cure. And as unworthily he served Dr. Welden, in whose house he so long lay sick, to the great trouble and disorder of his family, prejudice and spoiling of his goods.

By this means the cause could not come to a hearing, as by the order of the 28th of May, 1673, was directed. But, during this the Doctor's sickness, he was several times sent unto by his Grace of Canterbury and the Lord Chancellor, to know if he would resign. He declared that he would, so soon as there were indifferent persons nominated to arbitrate the differences between him and the parishioners, concerning arrears of tithes.

Thereupon, his parishioners presented the petition annexed, and had the order of reference under-written made thereupon:—

To his Grace the Lord Archbishop of Canterbury, and the Right Honourable the Lord High Chancellor of England, the humble Petition of the Inhabitants of Croydon,

HUMBLY SHEWETH,

That your petitioners most thankfully acknowledge your Lordships' great favour in giving them hopes of the removal of Dr. Cleiver, their vicar, and placing amongst them Mr. Hescott, who hath been two Sundays with them, and preached, and the whole parish in general satisfied with him, and are ready to make good their former proposals, under hands, for making a settled maintenance of 160*l.* *per annum*, to be secured and paid in such manner as therein was proposed. Sunday, the 22nd of June instant, the church doors were shut up, there being no person to preach or read prayers; which was a great discouragement to the parish. That Dr. Cleiver, who (as your petitioners are informed) had resigned his vicarage (or promised on Friday last was seven-night so to have done), doth now refuse to make such surrender, until by your Lordships two gentlemen be nominated and appointed to arbitrate the matter in difference between him and some of the parishioners, concerning his tithes in arrears; to which your petitioners readily consent.

That, by the order of his Majesty in council, upon the last petition presented by your petitioners, it was referred to your Lordships to end this business, if you could, or otherwise the cause upon the said petition to have been heard at that board the 11th of June last past. That your petitioners have none to preach to them, or christen, marry, or bury, Mr. Hescott being gone into the country.

The premises considered, they most humbly implore your Lordships' favour to nominate and appoint, under your hands, two persons to arbitrate the difference aforesaid, and desire them to meet and end the same, and that Dr. Cleiver may give his final determination whether he will surrender or not; and the cause thereupon to be set down to be heard in council; and in the meantime, that you will be pleased to appoint some good man to preach amongst them.

And, as in duty bound, they shall pray.

July 24, 1673.

We do nominate and appoint Sir Adam Brown, Baronet, and Sir William Haward, to be the persons to arbitrate the matter in difference about tithes in arrears between Dr. Cleiver and the inhabitants of Croydon, and do desire that they will undertake the trouble of ending these differences between them.

Gilb. Cant.
Shaftsbury, C.

Whereupon, Sir Adam Brown and Sir William Haward appointed to meet at Croydon, the 5th of August, 1673; where they that day, and on the 6th, 7th, and 28th of August, and on the 2nd and 9th of September, met to hear and compose the matters in difference between the said vicar and his parishioners.

And they heard and determined about two hundred and fifty differences, and made their report in express words as followeth:—

To his Grace the Lord Archbishop of Canterbury, and the Right Honourable the Lord High Chancellor of England.

May it please your Lordships,

In pursuance of your Lordships' reference unto us made, upon the petition of the inhabitants of the town of Croydon, in the county of Surrey, dated the 24th of June last, we have been several days upon the place (to wit), on the 5th, 6th, 7th, and 25th of August last, and the 2nd and 9th of September instant, to endeavour to compose the matters in difference betwixt Dr. William Cleiver and those of his parish with whom he had any controversy. On the 5th of August, the Doctor, after some delay, came to us to the George Inn, where many of the parishioners being present, we examined as many of their cases as we could before ten of the clock at night, when the Doctor would stay no longer, though one hundred of people were waiting; so we adjourned to eight o'clock the next morning, when the Doctor refused to come to us, alleging indisposition of body. Whereupon, resolving to proceed as far as we could, we went to his house, and continued sitting from eight in the morning till half an hour past twelve at night, without any intermission to eat; in which time we were sometimes told by the Doctor, that we needed not to trouble ourselves any farther, for, that he and those that were in arrears to him (and which we had not before determined) should agree. But, upon the importunity of the poor, we stayed and heard all persons that the Doctor then complained against, and determined every case then before us; so that we thought to have made our report. But afterwards, accidentally riding through the town, and hearing fresh complaints, upon the desire of many of the inhabitants, we appointed to meet again on the 25th of August last; when accordingly we came to the town, and sent for the Doctor, who refused to come to us. .Whereupon, we again went down to his house, but were so delayed by him, that we could not in the forenoon dispatch above eight or nine persons, he refusing to let those be called that were present, and ready to pay money, and calling upon others that were not there, and who only owed him for christenings and offerings. And, having sat till one of the clock, we went to the George to dinner, desiring the Doctor's company with us, offered him the convenience of a coach, because of his lameness, and left the same to bring him; went ourselves on foot, the which we rather did, because his own house was inconvenient, and we were much incommoded there by the smallness of the room and the heat occasioned by the crowd of people; but the Doctor would not come to us. So that, after sending for him several times, and he refusing to come, we writ to him, and appointed him to come to us on the 2nd of September instant;

expressly desiring him to prepare us a list of the persons' names that owed him any money, and how much each owed him, and for what; that so we might lose no more time in searching his books and papers for all the men that came before us, which much injured our proceedings; which done, we dismissed those that attended, who lost that daye's work, and appointed them to attend on the 2nd of September. But, on the 1st of September, the said Doctor prevailed (upon pretence that he was ill by preaching the Sunday before), to defer the meeting till the 9th of September, when he promised to come to the George to us; and that, in the mean time, the parishioners should have notice thereof. Nevertheless, we, loth to disappoint them, came to Croydon the 2nd of September, and found many of the inhabitants waiting, having had no timely notice; and therefore sent to the Doctor, desiring him to come to us and dispatch them, (being informed he was well, and went abroad that day); but, though we sent often down, he refused to come; and so we were forced to dismiss them, with the loss of that day also, which much troubled us, appointing them to appear again the 9th of September; when we again attended, and sent for the Doctor, who at first refused to come, pretending that he had not promised so to do; whereby he spent great part of our time, insomuch that we could do nothing till after dinner, when we sat, and the inhabitants being present, we demanded the list aforesaid, but could not prevail for the same; so that most part of our time (as formerly) was spent in his turning over his books. Nevertheless, we sat till ten at night, and dispatched all that were present, unless some few which the Doctor would not stay to end, though he continued longer in the room than we believe would have dispatched them; nor would he agree, though we desired it, to appear the next morning to end the same; so that they lost their labour. And we do humbly certify, that, having examined about two hundred and fifty several cases, we did find, that the occasion of the difference between the Doctor and them was the unreasonable demands for tithes, and other undue impositions he had endeavoured to impose upon most of them, being the poorest sort of inhabitants; who, being illiterate, were, by undue means, drawn into exorbitant agreements, penal bills, or judgments, by arresting and keeping them in durance, till they complied with him therein; whereby many families were put to very great expense, and thereby much impoverished some of them, totally ruined others, forced them to run from their houses, and leave their wives and children to the parish; by which means we conceive that he himself, in some measure, hath been the occasion of the non-payment of what was his just dues, which otherwise might probably have been had, with the affections of the people. And for ground of this our belief, we take leave further to inform your Lordships, that every person that appeared before us, and with whom the Doctor had any contest, most willingly submitted to whatever we should determine between them; whereupon we proceeded to examine his demands, and, upon deliberate consideration of what was alleged on each side, found them very extravagant, and so fitting to be submitted unto; therefore moderated the same, giving him, in some cases, but the twentieth, in others the twelfth, tenth, sixth, or fourth part of what he demanded; which, in many cases, was more than they ought to have paid. Nevertheless, they most readily submitted thereunto, and all of them paid the same, in hopes of the Doctor's being removed, so that they may not further be troubled with him; the which he seemed to insinuate his readiness unto, and was the great motive that in many cases invited us to do what we did, excepting some poor alms-men, from whom he had gotten judgments for several sums of money, who being to pay a part thereof, whereupon, finding little or no ground for the said judgments, we reduced the sums demanded to a third part, and the parish undertook to satisfy the same, with which the Doctor rested satisfied; and also two or three small sums, not exceeding five pounds in the whole, for payment whereof we have given some short time. And we did determine all the cases that came before us, except one of Christopher Joyner, deceased, of whose executors he demanded 10*l. per annum*, for ten years' arrears of tithes, when as he had not above 20*l. per annum* in the parish; and Sir Purbeck Temple and Thomas Bower's cases, the difference between whom had been before submitted to reference, and awards made therein. Nevertheless, they submitted to our determination; and Sir Purbeck Temple tendered the money according to his award, which would not be accepted; and excepting also the case of Edward Harvy, the which had also been submitted unto reference, and one award made therein (though he be not able to pay any part of the money); which cases we were forced to leave, as we found the Doctor refusing to refer himself to us therein. And, having thus proceeded (being weary with these delays), we left the said parishioners, who most humbly and unanimously implore your good Lordships' favour to remove from amongst them the said Doctor, and that a good man may be placed amongst them. In which suit we also humbly join to your honours, as a thing which, for the reasons aforesaid, we do judge very convenient; all which, nevertheless, we humbly submit to your Lordships' judgments, and remain,

My Lords,
Your Lordships' most humble servants,
ADAM BROWN.
WILLIAM HAWARD.

Croydon, September 10, 1673.

This report is not the tenth part so bad as the matter appeared against the Doctor; the particular cases decided are too long to insert; but if they were, it would not be possible for any man to believe there could be so many horrid oppressions and frauds put upon poor people by any clergyman living. Of this report there were two copies signed; the one of them the referees presented to my Lord's Grace of Canterbury, the other to the late Lord Chancellor; who thereupon promised to do what in them lay to remove the said Doctor. Accordingly he was sent for, promised his Grace that he would resign, and came to the Lord Chancellor, and told him the same also; and that, whenever his Lordship should desire the same, he would do it; adding, that his Grace commanded him to tell his Lordship that he would present to Croydon such a person as his Lordship would appoint. And upon this consideration, the Lord Chancellor promised the said Cleiver to provide for him some other living, bidding him rely upon his honour for it, assuring him he should have the first convenient living that fell in his gift, or to that effect; whereupon the said Doctor promised to conform to his Lordship's desires, but neglected to do the same, shuffling and endeavouring to put tricks upon their Lordships. This necessitated the parishioners to petition his Majesty and council once more; therein setting forth what the referees had done, annexing their report to the said petition, shewing the necessity of the Doctor's being removed, or the town's being ruined; therefore humbly implored his removal, or that they might have his Majesty's royal leave for to insert a clause into the act of Parliament then drawing, to settle the 160*l.* *per annum* on the vicar; to make the said Doctor incapable of any preferment in church or state.

This petition being read, the King's most excellent Majesty present, the parishioners were called in, and told by the Lord Chancellor that their petition had been read, and that his Majesty's royal pleasure was, that in case the said Doctor did not surrender before the bills designed to be brought into Parliament for settling the maintenance aforesaid should be presented to the Parliament, that then they should put in such a clause into the said bill, wishing the parish to hasten the said bill. This the Lord Chancellor declared to them in his Majesty's presence. Besides this, the order following was made:—

At the Court at Whitehall, October **15, 1673.—Present, the King's most excellent Majesty in council.**

Upon reading the petition of the inhabitants of Croydon, complaining against Dr. William Cleiver, their vicar, and praying that he may be removed, for several reasons set forth in the petition; his Majesty in council hath this day declared, that when the inhabitants of the said parish of Croydon shall provide to settle 160*l.* a year as a maintenance for the vicar of the said parish, according to their proposal in the said petition, in order to the passing of an act to confirm the same for ever on the vicarage of the said parish, then his Majesty will give effectual order for removing the present incumbent, Dr. Cleiver.

ROBERT SOUTHWELL.

Upon this, the parishioners immediately went home, and published in the market, and at the church, his Majesty's gracious answer; desiring all the inhabitants to meet at the vestry to sign a declaration testifying their consents to the passing of the bill aforesaid; and accordingly most of them did meet, signed and sealed such certificates of their consents, and their humble petition to both houses of Parliament to pass the said bill, and empowering a solicitor to prosecute the passing thereof.

This done, a bill was drawn, and such a clause as aforesaid was contained therein; when drawn, the same was copied, and one copy carried and left with his Grace of Canterbury; the other with the then Lord Chancellor: They having perused the same, my Lord of Canterbury referred the amendment thereof to Mr. Phillips of the Inner Temple, his Grace's counsel, and the Lord Chancellor to Mr. Attorney Montague; both were attended, and they made their several amendments: with which the same was fairly copied, and put into Sir Adam Brown's hand, to present in October last; but the prorogation of Parliament prevented the same.

All this time the said Doctor had notice of it; knew of the clause; promised to surrender, so it might be left out; went to both his Grace of Canterbury, and from him to the Lord Chancellor, frequently declaring the same. And my Lord Chancellor offered him another living, which he kept void for him; and so in truth it was, till the very day his Majesty was pleased to send for the seal from his Lordship. But the Doctor, resolving to persist in his villainies to ruin the parish of Croydon, (as he frequently declares he will do before he leaves it), refuses to surrender, but continues to go on in horrid oppressions and vexations, commencing suits against his parishioners, without colour of cause, to their unspeakable damage.

Under these sad oppressions, the poor parish having lain these thirteen years languishing, they now become humble supplicants to the Parliament of England, to enable them by an act to give such maintenance to a succeeding minister as may be an encouragement to a sober, learned, orthodox, and peaceable man to come and settle amongst them; to do the church that right as to remove so wicked and scandalous a person out of it; and, for the honour and vindication of the religion of the Church of England, to make him for ever incapable of serving in the church again; than which no greater advantage can be done to the Church of England at this time.

www.ingramcontent.com/pod-product-compliance
Lightning Source LLC
LaVergne TN
LVHW012206040326
832903LV00003B/160

* 9 7 8 1 0 1 7 6 3 4 5 9 4 *